Henry Gardner Landis

How to Use the Forceps

With an Introductory Account of the Female Pelvis and of the Mechanism of

Delivery

Henry Gardner Landis

How to Use the Forceps
With an Introductory Account of the Female Pelvis and of the Mechanism of Delivery

ISBN/EAN: 9783337779443

Printed in Europe, USA, Canada, Australia, Japan

Cover: Foto ©berggeist007 / pixelio.de

More available books at **www.hansebooks.com**

How to Use the Forceps.

WITH AN

INTRODUCTORY ACCOUNT

OF THE

FEMALE PELVIS

AND OF THE

Mechanism of Delivery.

BY

HENRY G. LANDIS, A.M., M.D.,

PROFESSOR OF OBSTETRICS AND DISEASES OF WOMEN AND CHILDREN
IN STARLING MEDICAL COLLEGE.

ILLUSTRATED.

NEW YORK:
E. B. TREAT, PUBLISHER, 757 BROADWAY.
FOR SALE BY MEDICAL BOOKSELLERS.

1880.

TO

A MASTER OF THE OBSTETRIC ART

AND

AN ESTEEMED FRIEND,

Ellwood Wilson, M.D.,

THESE PAGES ARE

RESPECTFULLY INSCRIBED.

CONTENTS.

PREFACE.

THE views herein set forth of the Anatomy of the Pelvis were imperfectly outlined in an article published in *The American Journal of the Medical Sciences* for April, 1876. Further study and experience in teaching have led to their expansion into what is now, I trust, a more exact and intelligible statement. The practical deductions which arise from them are given with as much conciseness as possible.

It has not been thought necessary to present an array of authorities and opinions of others as to the manner of using the Forceps when the standpoint of observation was obviously different. With this disclaimer of improperly ignoring the labors of others in this field, these pages are submitted to the profession for the test of an enlarged experience.

H. G. LANDIS.

COLUMBUS, O., Sept., 1880.

INTRODUCTION.

THE right use of the obstetrical forceps demands a thorough knowledge of four things : First, of the instrument itself, its form, design, and capabilities ; second, of the place into which it is to be introduced, viz., the maternal passages, their form, direction, and mutual relations ; third, of the body upon which they are to be applied, viz., the child's head, its form, consistence, and tolerance of manipulation ; fourth, of the normal mechanism of labor, or the manner in which the child should be delivered by the natural powers—for the forceps are not a foreign and unnatural resort, like the Cæsarean section, but are intended to assist, supplement, and conform to the course naturally observed in labor. The great diversity in the shape and design of forceps now in use, and the vague and conflicting opinions as to the manner of their employment are a sufficient evidence that an exact and scientific basis has not yet been reached or, if known at all, that it has not been well and generally understood. A study of the mechanism of labor *de novo*, will be, then, the first requisite for a proper understanding of any artificial aid intended to assist or replace that mechanism. I shall take for granted a preliminary acquaintance with the superficial anatomy of the pelvic bones.

PART I.

THE MECHANISM OF LABOR.

SECTION I.

THE ANATOMY OF THE PELVIS.

THE mechanism of labor is concerned with three things. 1. A body to be propelled. 2. A tube or channel through which it is propelled. 3. The forces which accomplish and regulate the propulsion.

The first is the child, and chiefly the child's head, which alone offers much resistance. The second is contained mainly in the pelvis. The third is mainly of muscular origin. The relations which these several factors bear to each other, and especially those subsisting between the first and second, constitute the most important part of the study of this mechanism. Neither of these can be profitably studied apart from the other except in so far as they may present conditions alien to the mere fact of delivery. As a starting-point we may take the most permanent factor, the pelvis.

The female pelvis has three uses :

I. It serves to contain and protect certain vessels and viscera.

II. Being placed at the end of the vertebral column it is designed to support the weight of the body, transmitting it to the femora in the erect position and to the ischiatic tuberosities in the sitting posture.

III. It is modified to allow and direct the passage of the child through it during labor, and is the principal constituent of the parturient canal. The first use is obvious, and is not relevant in this connection. The second is not entirely relevant, and may be dismissed with this brief formulation, which the practically minded reader may omit.

1. The pelvis is made up, first, of two beams, the *sacro-iliac*, extending laterally from the base of the vertebral column to the acetabulum of either side and thus distributing the weight of the body to the femora in the erect posture.

2. These lateral beams are continuous with a third beam, the *pubic*, placed transversely, and in front, which regulates the interval between them.

3. These three beams in the adult female are arched outwardly to provide room for the parturient act, and are so situated as to form a complete bony rim at the beginning of the pelvis.

4. From the under side of this rim two other arched beams spring, the *ilio-sciatic*, one on each side and posteriorly, which end in the ischial tuberosities, to which they transmit the weight of the body in the sitting posture.

5. A sixth arched beam, the *sub-pubic*, is placed under the bony rim in front, which also has its extremities in the ischial tuberosity of either side.

6. The upper bony rim is amplified into a tube by the presence of these secondary arched beams on the front and sides, and by the extension of the sacrum and coccyx behind.

Thus we see that the pelvic tube is not entirely designed as a parturient canal, but that a structure having other uses has been modified for this secondary purpose. The extent of the modification can be seen by comparing the male and infantile pelves with

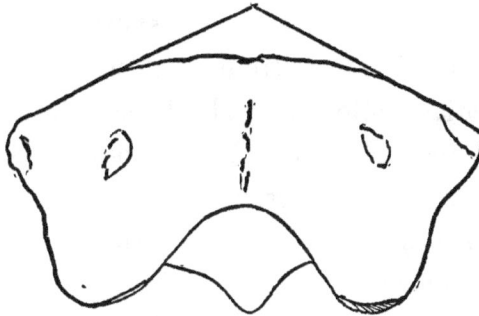

Fig. 1.—Outlined from Hodge.

that of the adult female, the beams of the former being nearly straight, while those of the female are greatly arched. And if it is modified for the sake of the child, we may expect to find a correspondence between the shape of the pelvis and the shape of the child. Before making the comparison, we will notice that the wings of the ilium and sacrum are concerned only with the first and second uses of the pelvis, being buttresses of the arched beams and guards of the viscera against external violence. The obstetrical relations of the pelvis begin with the bony rim before mentioned. We may therefore remove these wings as a preliminary to our study. When the sacral and iliac wings, or "false pelvis," are removed, the pelvis presents the appearance shown in Fig. 1,

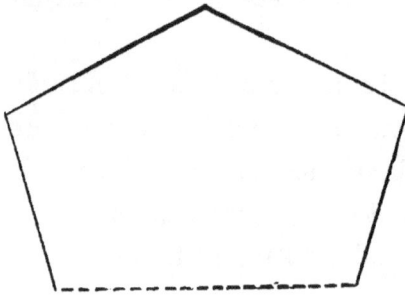

FIG. 2.

when viewed from in front. If we then make a perpendicular section through the acetabula we shall find that the pelvic tube has an outline similar to that shown in the diagram Fig. 2. It is therefore wider above than below, which is the first important fact to remember. This does not give us a complete idea of the tube, for the sacrum which forms its posterior wall is markedly curved. We must therefore make another perpendicular section at right angles to the former one, which will give us such an outline as is

FIG. 3.

shown in Fig. 3. By combining these mentally, for

obviously no pictorial representation can show them at once, we will begin to have an approximate idea of the shape of the pelvic tube. But we would, if we stopped here, have an idea that it resembled a funnel bent upon itself, and would fail to have any explanation why the child in labor does not at once drop to the bottom, since the top of the funnel is so much more capacious than the lower end. From these two sections we learn

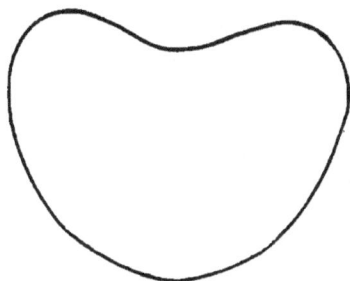

FIG. 4.—THE PELVIC INLET.   FIG. 5.—THE PELVIC OUTLET.

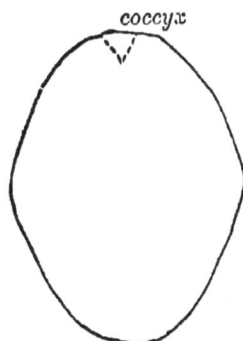

only the direction of the tube ; its calibre must be determined by looking into and through it. Its beginning or inlet is found to have the shape indicated in Fig. 4 ; its outlet, that shown in Fig. 5, a remarkable difference. These four figures (2, 3, 4, 5) show the pelvis from in front, from the side, the inlet and outlet, and must be held in mind while we seek for a something to harmonize and explain them.

Beginning with the inlet, we find that its shape is often spoken of as an irregular oval, but when we analyze it we will find that it is beautifully regular in outline. The explanation of its shape must be sought, as said before, in the child, for which the pelvis has been modified. Clinical experience teaches us that the child's head is the part which offers the most resistance in delivery. Its great relative size and firm organization make it the most difficult part to be expelled. Also, it is usually in advance, and after its passage through the pelvis the rest of the body can readily follow. Next to the head the shoulders offer the largest outline. Only under exceptional and abnormal circumstances do any other parts of the child present any difficulty in passing through the pelvis. The natural manner for the child to enter the pelvis in labor, is with the top of the head in advance.

The middle circumference of the head is therefore applied to the brim or inlet at the beginning of labor. If a plane section be made through the middle of the head horizontally and at the level of the parietal eminences, it will be bounded by such an outline as is shown in Fig. 6, which is for all practical purposes an ellipse. As a matter of fact, if the head is partially flexed upon the breast, a

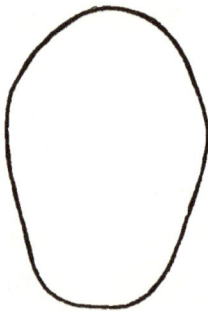

FIG. 6.—OUTLINE OF FŒTAL HEAD.

horizontal section made at the same level will be entirely elliptical. If we apply an ellipse cut out of

card-board and having such an outline, to the inlet
of the pelvis we will find that it completely coincides
on one side, and if reversed, to the opposite side, the
two outlines intersecting one another. This is shown
in Fig. 7, where the
dotted line A B finishes
the elliptical outline on
one side and the line
A C upon the other side.
The same ellipse applied
to the outlet entirely
corresponds to it, though

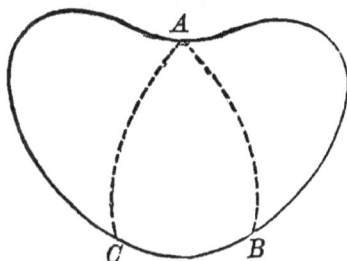

FIG. 7.

the outward flaring of the ischial tuberosities makes
this a little wider.

We may then say, tentatively, that the outline of
the inlet is compounded of two partially superim-
posed ellipses similar to the outline of the fœtal head
—while the outlet represents but one such outline.
The shoulders will throw more light upon the subject. They also have upon transverse section an elliptical outline almost identical with that of the head.
But the long diameter of the shoulders, i.e., their
breadth, is at right angles to the long diameter of
the head. Therefore, when these two ellipses are su-
perimposed, as happens practically when the shoulders follow the head through the pelvis, their outline
would present such an appearance as is shown in
Fig. 8.

This is not the whole truth. The foramen magnum, and therefore the occipital condyles, are not placed centrally in the base of the child's skull, but much nearer the posterior end of the head, especially

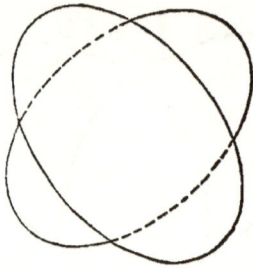

FIG. 8.

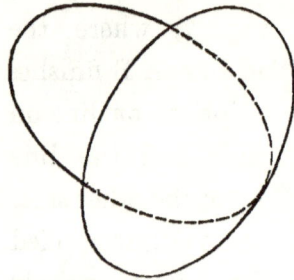

FIG. 9.

when the head is flexed. Therefore, Fig. 9 may be substituted for Fig. 8 as more exactly representing the facts. Now if the head is laterally flexed so as to bring one ear nearer to the corresponding shoulder than in the horizontal position, which also happens during the labor, these outlines would be superim-

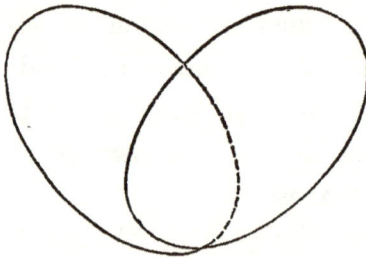

FIG. 10.

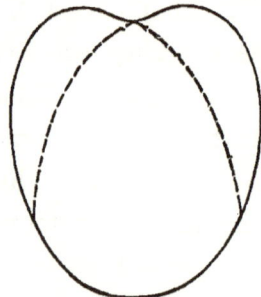

FIG. 11.

posed in the manner shown in Fig. 10, which also represents the outline of the pelvic inlet (Q. E. D.). The length of the pelvis is such that the shoulders may still remain in the upper part when the head is born. If it were not for some such provision the neck would be disagreeably twisted, by reason of the shoulders being compelled to follow the head through a passage calculated for the latter alone. Furthermore, transverse sections of the pelvic tube made at any point will show this double relation until we reach the outlet, where there is evidently but a single canal. Fig. 11 shows the outline of the canal a little above the outlet. By the time the shoulders have reached this point the head is born.

We may therefore infer that the pelvis is in reality a *double canal*, its two parts being partially fused at the beginning and entirely so

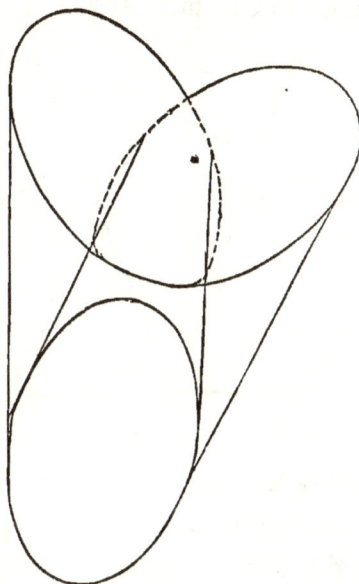

Fig. 12.

at the end, and may construct a theoretical diagram, Fig. 12, which will exhibit these facts. This will explain the appearance delineated in Fig. 2, for on adding dotted lines to represent the inner and invisible walls of these supposed parts, as in Fig. 13, we see why the pelvic inlet is wider than the outlet, and also learn the direction of the two canals.

These facts may be formulated as follows, before proceeding further, with such conclusions as may warrantably be drawn from them.

I. The pelvis contains two canals, partially separate at the beginning and identical at their termination.

II. These canals converge from above downwards, and are also mutually curved from before backwards,

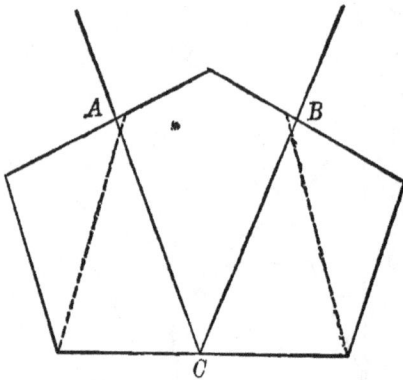

Fig. 13.

as indicated in Fig. 3. Their direction is therefore somewhat spiral.

III. The calibre of each canal is that of the fœtal head ; therefore the head may descend in either canal and will follow a spiral course in so doing. These canals may be called respectively the right and left canals ; the right being the one in which exclusively the right sacro iliac symphysis is found, and the left in which the left sacro iliac symphysis is found. Of these the right is somewhat the larger and is the one in which the head usually descends ; for which there are other reasons, as will be shown further on.

For purposes of description certain planes, axes, and diameters are to be considered, concerning which we will first state the views generally enter-

tained. Playfair says :* " By the planes of the pelvis are meant imaginary levels at any portion of its
circumference. If we were to cut out a piece of
card-board so as to fit the pelvic cavity, and place it
at the brim or elsewhere, it would represent the pelvic plane at that particular part, and it is obvious
that we may conceive as many planes as we desire."
Two such planes are of especial importance, those of
the inlet and outlet,
or, as they are also
termed, the superior
and inferior strait.
Hodge defines the
plane of the superior
strait as a surface
bounded by the cir-
cumference of the
strait which is marked
by the " inner margin
of the tuberosity or

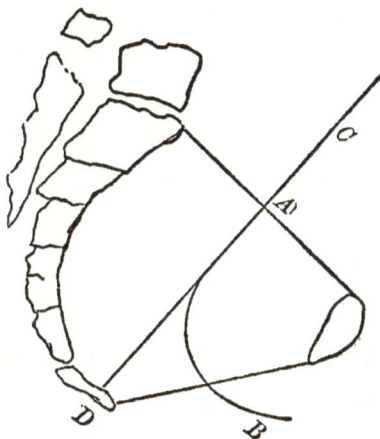

Fig. 14.

horizontal portion of the pubes on either side, by the
spinous process, the linea ilio-pectinea, and the inner
margin of the ala of the sacrum, and posteriorly by
the promontory of the sacrum." The axis of this
plane is a line drawn at right angles to it and the
combined axes of similar planes drawn at all levels of
the pelvic cavity, constitutes the axis of the pelvis,
which is supposed to indicate the course of the
child's head in delivery (see Fig. 14).

* System, p. 35–6.

By beginning wrong we generally end wrong.
By studying the pelvis only from antero-posterior sections we get only a partial knowledge of it. The
whole study of the mechanism of labor as given by
Hodge and his successors is vitiated by the inaccuracy involved in his description of the superior
strait. For to this succeeds, as a consequence, a
vagueness as to the true position of the head in labor
—which is a point of great practical importance, especially when we attempt to apply the forceps. To
begin on common ground, the plane of the inferior
strait is confessedly artificial and arbitrary. The
outlet is so irregular in its termination that no one
pretends to describe a plane passing through all the
points of its circumference. We draw a line from
the under edge of the symphysis pubis to the tip of
the coccyx, called the conjugate diameter of the outlet, and a plane passing transversely through this line
and limited by the calibre of the pelvis, we call the
plane of the inferior strait. Since this calibre is so
evidently the same or nearly so as that of the fœtal
head, and since we find clinically that the head
emerges from the outlet in a definite relation with
such a plane, we may retain it, but always admitting
its artificial character and boundary. The same
course has not been followed with the inlet. Hodge
gives no hint of compromise in fitting the plane of
the superior strait in its circumference, although
Fig. 1, outlined from his work, shows clearly enough

that no plane can pass through the points mentioned in his definition. And as unnoticed error, especially when sanctioned by high authority, has a great power of growth, it is not surprising to find Dr. Leishmann following with the statement that the various parts of the line bounding the superior strait "are in man alone on the same plane." As a matter of fact, the circumference of the inlet bounds two distinct planes, whose inclination to each other may be seen in Figs. 1 and 2 to be about at an angle of 150°.

If we cut out of card-board two ellipses similar in outline to the middle circumference of the fœtal head and apply them or attempt to apply them to the border of the pelvis on each side—in other words to the very points mentioned above by Hodge—we will find that they intersect one another in the median line, while accurately fitting the pelvis in other respects. We may call these planes respectively the initial plane of the right and left canal. Any number of similar planes may be drawn in each canal, which will have a less and less inclination to each other until at the inferior strait they will be identical with each other and with the plane of the inferior strait as above described. The axis of the initial plane of either canal is a line drawn at right angles to that plane, and indicates the direction of either canal at the beginning.

The *axis* of each canal will be a line extending from the centre of its initial plane centrally through

the canal to the centre of the plane of the inferior
strait.   This line will not have only the direction
shown in Fig. 13, but being curved from before backwards, in the manner of the central axis in Fig. 14,
will be spiral and therefore incapable of pictorial representation.   But there is upon the pelvic walls a
line on either side, which is as nearly as possible parallel to this axis, viz., the raised line extending from
each pectineal eminence on the ileo-pectineal line to
the ischial spiné of the same side.   As this is an important line from this circumstance, and from the
part it plays in the mechanism of labor, we may give
it a name and call it the *ilio-sciatic* line.

It will be convenient for descriptive purposes to
retain the so-called "plane of the superior strait,"
but for avoidance of confusion we may define it as
passing transversely through the conjugate diameter
(CD, Fig. 15) of the inlet and call it the *plane of the
conjugate diameter*.   Similar planes may be conceived
of as drawn at right angles to the general cavity of the
pelvis at any level, and to distinguish them from like
planes drawn in the right and left canals we may call
them *planes of the pelvic cavity*.   The plane of the
conjugate diameter is said to be inclined to the horizon at an angle of 60° when the woman is in the
erect posture, the face of the pubes looking almost
directly downward and the plane of the outlet backwards and downwards.   In the sitting posture, with
the pelvis resting on the tuberosities of the ischia, the

inclination of the plane of the conjugate diameter is about 45°, while the plane of the outlet is almost horizontal and looking directly downwards.

In the recumbent posture the plane of the conjugate diameter is almost equally inclined in an opposite direction from the last, the plane of the outlet being nearly vertical. In the semi-recumbent posture, which is supposed to be the characteristically American method of sitting, the plane of the conjugate diameter is level with the horizon, while that of the outlet looks downwards and forwards. The initial planes of the right and left canals have substantially the same inclination to the horizon as the plane of the conjugate diameter in these various positions, so far as the planes are considered in their anteroposterior direction. But they have also a lateral obliquity of about 15° from that of the conjugate diameter, which is made sufficiently evident by reference to the figures or better still, to the pelvis itself.

Certain diameters are usually described as existing in the inlet and outlet of the pelvis.

The principal ones in the inlet are the two oblique diameters and the conjugate. The oblique diameters are drawn from the sacro-iliac symphysis of either side to a point slightly in advance of the pectineal eminence of the opposite side (Meadows). In Fig. 15, AB represents the right oblique diameter (according to the German nomenclature), and EF, the left oblique. If we apply to the inlet a piece of

card-board cut after the pattern of the elliptical outline of the fœtal head, as delineated in Fig. 6, we
will see that the long diameter of such an ellipse corresponds with the oblique diameter of the canal in
which it is inserted, while the short diameter of the
ellipse lies in the line of the opposite oblique diameter.   These diameters are nearly or quite five inches
long in the normal pelvis,
and are longer than any
other which can be drawn
in the pelvic brim, except
in some cases the one ex-
tending directly across it
and known as the trans-
verse   diameter.   The
conjugate diameter CD is
drawn from the promontory of the sacrum to the
middle of the top of the symphysis pubis, and is the
shortest, being about four inches in the normal pelvis.   Two others should perhaps be mentioned here,
which are the ones drawn across the base of each
sacro-iliac arch and called the *sacro-cotyloid.*

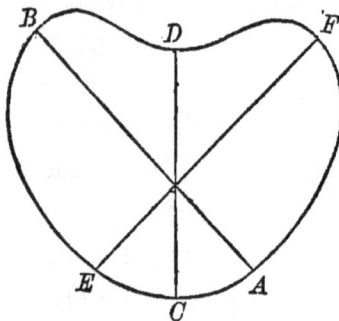

Apart from any consideration of the doubleness
of the pelvis, it is generally recognized that the head
will enter the inlet with the greatest economy of
space when its long or occipito-frontal diameter coincides with one of the oblique diameters of the pelvis, while its transverse or short diameter has an
equal amount of room in the opposite oblique diam-

FIG. 15.

eter. At the outlet the antero-posterior or conjugate diameter extends from the under edge of the
pubes to the tip of the coccyx, but a little reflection
shows that it is the representative of the upper
oblique diameters. Thus, if a rod be placed in the
inlet coincident with the right oblique diameter, and
its central point carried downwards in the axis of the
right canal, its posterior extremity will traverse a
line from the right sacro-iliac symphysis to the tip
of the coccyx, while its anterior end will follow a
similar line from a point in front of the left pectineal eminence to the centre of the under edge of the
symphysis pubis, and the rod will then lie in the
conjugate diameter of the outlet. For the rod,
substitute a fœtal head with its long or antero-posterior diameter in coincidence with the right oblique
diameter of the inlet and the correspondence of the
head to the right canal throughout will be entirely
manifest. The same may be affirmed of the left canal,
with a corresponding change of right to left, and so on.

The bony pelvis, with its ligamentous and muscular additions, does not comprise the whole of the
parturient canal, but at the inferior strait begins that
part of it which is made up only of the soft parts.
The latter is only temporarily fitted for this use, and
has no fixed calibre, axis, or diameters, which are
regulated by the size and shape of the fœtal head
and the direction taken by it. It is enough at present to conceive of it as an elastic tube through which

the head passes after being delivered from the pelvic
canal. Besides the soft parts at its termination the
uterus also may be said to form a part of the parturient canal, since the child in passing out of it must
have its original direction controlled to a great extent by the position of the uterus. The uterus, during labor, is not placed directly in the median line.

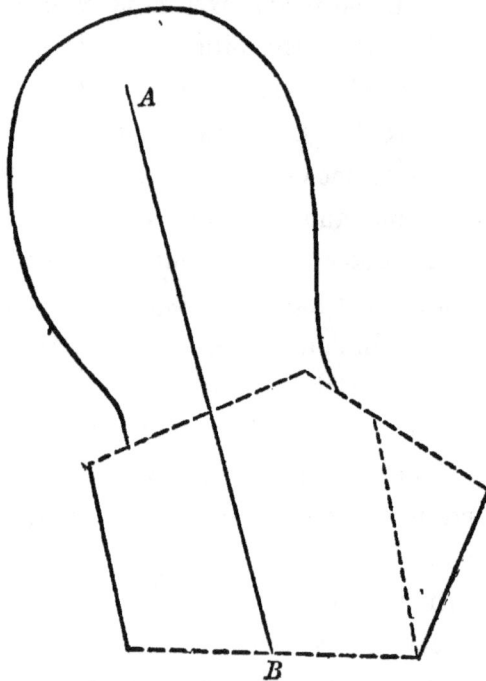

FIG. 16.

From various causes, among which the prominence
of the lumbar vertebræ is conspicuous, it is somewhat deflected towards one side or the other of the
median line, and in the majority of instances towards
the right side. Viewed laterally the womb appears

to be situated with its axis in the same line with that of the plane of the conjugate diameter. This is the statement usually made ; but when we perceive this obliquity we recognize that this cannot be, and that the axis of the uterus in labor is in the majority of instances continuous with the axis of the initial plane of the right canal (see Fig. 16).

If, then, the child is disposed in the womb with its long axis coincident with that of the womb, it will be situated in the most favorable manner for entering the right canal. And the same might be affirmed of the left canal if the womb was in a condition of left obliquity. Since it is rarely found in this condition, and since we find clinically that the head is found with similar infrequency in relation with the left canal, we derive additional proof of the doubleness of the pelvis and of the existence of such planes and axes as have already been described.

SECTION II.

THE forces concerned in the mechanism are of two kinds, propulsive and guiding. The former are furnished by the contraction of the uterine muscular fibres and by the voluntary and semi-voluntary contractions of the abdominal muscles, but not exclusively. The force of the uterine contractions is communicated to the vertebral column of the chlid and acts primarily in the long axis of the womb. They therefore tend to propel the child in the axis of the initial plane of the right canal in the majority of instances ; or when the womb has a left obliquity in that of the opposite canal. So far as the uterus is concerned, therefore, the child tends to move through the pelvis in the direction of the line AB, in Fig. 16. The abdominal muscles transmit force in the same manner to the child, but in the median line when they act uniformly. Hence they are well designed to propel the child after it has reached the inferior strait and has finished its oblique course in the right or left canal. And as a clinical fact, we find that the action of the abdominal muscles is not, as a rule, called into effect until the head has attained this stage. But from the inclination of the pelvis to the vertebral column, each of these forces, the uterine and abdominal, tends to propel the child to points behind the centre of the plane of the outlet. The abdominal muscles acting primarily in the axis of the plane of the conjugate diameter, impel it towards the tip of the coccyx. The uterine force tends to impel it to the same point ; perhaps a little to one side, but as far back.

These tendencies are modified by the directive or guiding forces reflected from the sides of the canal, which being spiral or screw-like in shape consist essentially of continuous inclined planes. From the pelvic outlet to the vaginal outlet the head follows a very different course, emerging from the latter in a direction which forms an acute angle with the produced axis of the plane of the conjugate diameter. It is obvious that the same force cannot act in two directions, one of which is almost the reverse of the other. There must be, then, a new force beyond the pelvic outlet acting in a different direction. This we find in the perineum.

The superficial or anatomical perineum is the space bounded between the posterior vaginal commissure, the anus, and the ischial tuberosities. The deeper structures of this area consist of certain muscles and fibrous tissue, and most important of all, the perineal body. They are placed in front of or

opposite to the pelvic outlet, constituting the floor of
the pelvis.   The mechanical action of the whole
structure may be studied in that of its principal
part.   The perineal body is a stout fibrous band extending from one tuber ischii to the other.   It is
made of elastic fibrous tissue, and both for strength
and elasticity is comparable to no other tissue of the

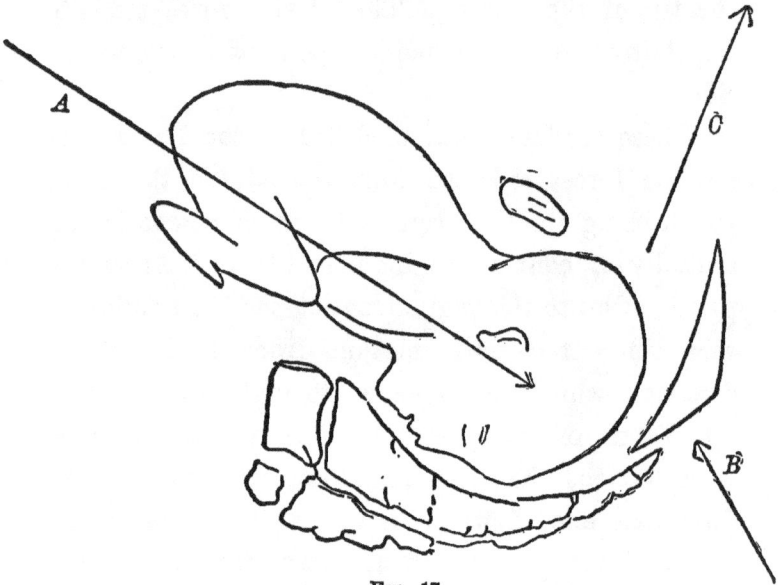

Fig. 17.

body unless perhaps the ligamentum nuchæ.   On section it appears wedge-shaped, being inserted between
the vagina and rectum at their termination, with the
edge directed upwards.   It extends backwards as far
as the coccyx—being covered and supplemented by
sundry muscles of more or less importance in this
connection, but having substantially the same mechanical purpose in labor.

When the combined propulsive and directing forces have brought the head to and nearly through the pelvic outlet, it is met by the opposing force existing in the elastic resistance of the perineal body assisted by the associated structures of the pelvic floor. The latter force acts in a direction nearly opposite to the former, and the head is, therefore, directed forward in the line of the resultant of the two forces. In Fig. 17 the arrow A represents the direction in which the original forces bring the head upon the perineum : B will represent the line of the direction impressed upon the head by the perineal force alone, and C will show the resultant of the two. The important practical bearing of this will be noted in the proper place. Another force may assist in delivery, viz., gravity. The amount of force employed in truly normal labor is not great.

The following, from J. Matthews Duncan (Researches, p. 319), will suffice to illustrate this point : " If we regard the figure of four pounds given by Poppel as equal to the power exerted in the easiest labor he has observed, or the corresponding figure of six pounds, according to my calculations, and keep in mind that the average weight of the adult fœtus exceeds either of these weights, we are led to the conclusion that in the easiest labors almost no resistance is encountered by the child ; that it glides into the world propelled by the smallest force capable of doing so ; that with the mother in a favorable position, the weight of the child is enough to bring it into the world—a result which many clinical facts at least appear to confirm." The same author says also : " Having had extensive and varied experience in the use of forceps in difficult labors, and having also made some rough experiments with the dynamometer, to ascertain the power I have applied by the instrument, I regard M. Joulin's estimate of a hundred-weight, as the maximum force of the parturient function, as too high- I do not deny that in very rare cases such a force may possibly be produced, but I am sure that it is nearer the truth to estimate the maximum expulsive power of labor (including the uterine contractions with the assistant expulsive efforts) as not exceed. ing eighty pounds."   In this opinion I entirely agree, believing that the extreme efforts made in some cases with the forceps are due to a misapprehension of the proper direction of force, rather than to any need for such an amount of force.

SECTION III.

THE child for whose sake all this machinery is ordered is, when packed in the womb, of ovoid shape. At one end is the head, at the other the similarly rounded breech. Like an egg, it is natural for it to pass through the pelvis endwise, with one end or the other in advance. As the head is freely movable upon the neck and capable of considerable extension, either the top of the head or the face may be in advance. The child may also attempt to enter the pelvis crosswise or transversely. We have then four distinct methods of entrance. The part in advance at the beginning of labor is called the *presenting* part, and the area of this part inclosed by the pelvic circumference is technically called the *presentation*.

There are therefore four distinct presentations : I. of the top of the head, or vertex ; II. of the Face ; III. of the Breech, and IV. Transverse.

The vertex presents in at least ninety per cent. of all labors and is evidently the natural presentation and the one for which the pelvis is specially designed. I shall, therefore, confine my remarks to this presentation, with a brief account of the second, since the others are neither strictly normal nor suited for the application of the forceps. Let us first refresh our memories with some topographical anatomy. Upon the top of the fœtal cranium appears a suture, extending directly antero-posteriorly between the parietal bones, called the *sagittal* suture. At its posterior limit is a triangular membranous interval called the *posterior fontanelle* or bregma. At its anterior limit is a similar quadrilateral interval called the *anterior fontanelle*. From the posterior fontanelle a suture extends on either side, joining the occipital to the parietal bones of either side, the two being collectively called the *lambdoidal* suture. From the anterior fontanelle a suture extends in front which is practically a continuation of the sagittal, called the *bi-frontal* suture. At right angles to it proceed from each side of the anterior fontanelle, sutures joining the anterior border of the parietal to the frontal bone, which are together known as the *coronal* suture. These fontanelles and sutures are of variable size, being sometimes large and distinctly recognizable, in other cases small and indistinct. They are also more or less obscured by the covering of the hairy scalp.

During labor the sutures themselves are not so apt to be felt as the overlapping edge of bone which results from their approximation. Between the limbs of the lambdoidal suture, an inch or a trifle more from the posterior fontanelle, is the occipital protuberance, a prominent and useful landmark in ascertaining the position of the head at times. Still more useful are the parietal protuberances situated nearly in the centre of each parietal bone. The frontal eminences similarly situated in the frontal bones are hardly to be felt except in face presentations or during the perineal stage of labor, and are not usually of diagnostic importance. The ears are scarcely ever within reach, still less the mastoid prominences. Not infrequently the small fontanelles at the postero-inferior angles of the parietal bones are within reach and are to be recognized by the extension into them of the lambdoidal suture. Certain planes and diameters are important for purposes of description. When the head is placed in a horizontal position *quoad* the body in the erect posture, a plane drawn transversely through the occipital and parietal protuberances will present an elliptical outline as already delineated in Fig. 6. This plane may be called, from its long diameter, the *occipito-frontal*, the latter line extending from the occipital protuberance to a point, in the bi-frontal suture. The transverse diameter is drawn from one parietal protuberance to the other, and is called the *bi-parietal*. It will be observed that this outline is not a perfect ellipse, the transverse diameter being behind the centre, though this irregularity is less

marked in the living head than in the dried skull.
The *occipito-frontal* diameter measures on an average
a little more than four inches in length, the bi-parietal about three and a half inches.

If the head is partially flexed, a similar transverse plane passing through the parietal protuberances will extend through the occipital ridge, or nape
of the neck, and the apex of the forehead, and may
be called the plane of *demi-flexion*. Its outline will
be almost exactly elliptical ; the transverse diameter
being the same as in the preceding plane, viz., the
bi-parietal, and its long diameter, the *cervico-frontal*,
will be a little less than four inches. If the head is
completely flexed, as when the chin rests upon the
breast, a similar plane drawn through the parietal
protuberances will pass through nearly the same
point posteriorly as in the last plane, viz., the nape
of the neck a little below the occipital ridge ; and its
anterior limit will be in the posterior margin of the
anterior fontanelle. Its outline will be nearly circular, for while the transverse diameter is still the bi-
parietal, its long diameter, the cervico-bregmatic, is
also three and a half inches long. This plane may
be called the *plane of complete flexion*. Its circular
outline is an important fact to bear in mind.

The conclusion is now apparent that flexion of
the head reduces the outline which it presents to the
pelvic passages. It is of interest to note the relative
position of the fontanelles during these changes.

When the occipito-frontal plane is horizontal the anterior fontanelle appears nearly in the centre of the elliptical area above the plane, while the posterior fontanelle is very near its posterior margin. When the head is in demi-flexion the fontanelles appear very nearly in the foci of the ellipse, and in complete flexion the posterior fontanelle occupies the centre of the circular area presented, while its fellow has disappeared from view in front.

The outline of the fœtal head is still further capable of being diminished by the overlapping of the' parietal bones, either by compression due to the small size of the pelvic canal, or artificially by the forceps. The bi-parietal diameter can be lessened from a half inch to a full inch by such compression. Not only can these diameters be compressed and shortened by these agencies, but the entire shape of the head may be changed by a process of moulding during the process of expulsion.

The face of the child offers little to detain us at this point. It is also of rather elliptical outline, having a long diameter, the *fronto-mental*, which extends from the chin to the top of the forehead ; and a transverse diameter, the *bi-malar*, which extends from one malar bone to the other. But these are so much smaller than the diameters which lie behind them in the head, that the face evidently offers no difficulties *per se* in delivery. The real difficulties are due to the manner in which the bulkier posterior portion of the head and the body are made to enter the pelvis when the face presents, and this can be better described in connection with the mechanism of labor in this presentation.

The *body* of the child exhibits upon transverse section an elliptical outline in its entire extent, and especially at the level of the shoulders and breech. As has already been noted, the long diameters of such sections are at right angles to the long diameters of similar sections of the head, and from the fact that the foramen magnum is situated behind the centre of the head, the body tends to follow the head a little behind the central axis of the head.

SECTION IV.

THE MECHANISM OF DELIVERY.

SINCE there is an evident correspondence between the pelvic canals and the head in their outline, it is a natural inference that the occipito-frontal plane of the head may enter either the right or left canal, and in two ways : with its occipital extremity either in front or behind. Clinical observation is usually in advance of theoretical knowledge, as is conspicuously shown by the fact that all recent writers agree in admitting but four positions of the vertex. And yet, while the pelvic brim is considered as having an " irregularly oval outline," there is no obvious objection to the eight positions of the earlier authorities, or indeed to any number whatever. It is only when we find that it is of singularly regular outline, by analyzing it, that we are compelled to see a theoretical reason for the already clinically observed fact.

I. THE VORTEX.

The nomenclature of these positions is founded on the position of the occiput, which will be situated on one side or the other and in front or behind. They are as follows :

1. Left occipito-anterior.
2. Right occipito-anterior.
3. Right occipito-posterior.
4. Left occipito-posterior.

1. THE FIRST OR LEFT OCCIPITO-ANTERIOR PO-
SITION (L. O. A.) is the most frequent, occurring in
at least seventy per cent. of all positions of the vertex. The reasons for its prevalence are to be found
in several combined causes. The folded-up attitude
of the child *in utero* requires that its back shall be
turned towards the mother's front. The prominence
of the vertebral column and more especially the sacral promontory, will determine the position of the
occiput on one side or other of the median line.
Since the long axis of the child is correspondent
with that of the uterus, its head is placed directly
over the initial plane of the right canal, owing to the
usual right obliquity of the uterus. Also the right
canal is actually a little larger than the left, and the
latter contains under the left sacro-iliac arch the rectum, which still further diminishes its size. This is
therefore the most natural and favorable of all the
positions.

At the beginning of labor the head in this position is placed with its occipito-frontal plane coincident with the initial plane of the right canal. The
occipital protuberance is opposite a point in front of
the left acetabulum ; the bi-frontal suture is in **front**

of the right sacro-iliac symphysis, and the right parietal protuberance is opposite a point over the right obturator foramen towards its inner edge. The left parietal protuberance is not opposed to any point of the pelvic circumference, but is in the free space in front of the left sacro-iliac symphysis. The occipito-frontal diameter is therefore coincident with the right oblique diameter of the pelvis. The head is obliquely situated with reference to the plane of the conjugate diameter, one side of the head being below and the other above that level. This fact was first noticed by Naegele, but stated too generally, since this obliquity frequently and indeed usually disappears in the succeeding stages. For as soon as the uterine efforts become at all effective, the head undergoes a compound movement by reason of which its synclitism with the initial plane of the right canal disappears, and therefore its obliquity to the plane of the conjugate diameter, while another head plane than the occipito-frontal is made to engage by means of flexion. The cause of this movement is to be found principally in the unequal resistance offered by the pelvic walls. The right parietal protuberance is directly applied to the anterior pelvic margin, while the left is entirely free, and the same may be said of the entire right and left sides of the head, the protuberances being cited merely as the more prominent parts. If the size of the head and the calibre of the right canal are at all equal and the fit is tight, the right side of the head will meet with considerable resistance to its onward motion communicated from the uterine forces, and will, therefore, be arrested, while the left side, being untrammelled, will descend. There will result a lateral flexion of the head, which will bring the occipito-frontal plane synclitic with the plane of the conjugate diameter.

If the abdominal muscles are called into action at this time, they will by their compression tend to force the uterus backwards and so deflect its axis as to still further impel the head against the anterior pelvic walls, which will also assist in bringing about this lateral flexion and the resultant synclitism. This synclitism is in reality an obliquity of the fœtal planes to the transverse planes of the right canal, and continues throughout the further progress of the head, being indeed necessary when the head has reached the inferior strait. Before that point it does not invariably occur. The relative size of the head may be small, and it may continue in exact relations with the successive planes of the right canal throughout, as was practically the teaching of Naegele. But inasmuch as the head is usually large enough to offer an appreciable amount of resistance, the synclitism of the presenting plane of the head with the artificial planes of the pelvic cavity is the rule rather than the exception. This has led Cazeaux, Hodge, Leishmann, and others to entirely combat the obliquity of the head at any time, which is an error in the opposite direction, since it must originally exist from the manner in which the head enters the inlet.

A similar cause to that which determines the lateral flexion of the head brings about at the same time flexion proper, or the movement of the chin towards the breast. Although the right oblique diameter with which the occipito-frontal is coincident is five inches long in the bony pelvis, the soft parts so diminish the size of the canal that some lessening of the head outline, especially in its length, is usually necessary. The head may be regarded as a lever attached to the vertebral column as a fulcrum. The resistance which it encounters causes the anterior part of the head, which is the long arm of the lever, to be flexed towards the chest. Also, the occipital end of the head is in the anterior and roomy part of the pelvis, and thus more free to move than the frontal end, which is cramped by the narrower dimensions of the right sacro-iliac arch.

This flexion continues until the head presents an outline small enough to pass readily, which usually happens when the plane of demi-flexion has become coincident with the plane of the conjugate diameter, or, to speak more accurately with a plane parallel to the latter, but a little lower in the pelvis. If the plane of demi-flexion presents too large a circumference, flexion continues until complete. If then there still remains any disproportion between the head and pelvis, the force is exerted upon all the diameters of the head, which is diminished in size by a general compression. From this results what is known as the moulding of the head, which so rearranges its shape that its original outlines are entirely changed and it becomes cylindrical. This, if successful, is continued until the diameters of the head correspond to those of the pelvic canal. I believe that this head-moulding often occurs at an earlier stage from a failure of the head to properly undergo flexion, and that a thoroughly flexed head is rarely in need of any further diminution of its outline. Ordinarily the plane of demi-flexion will have a sufficiently small circumference, and the head is then ready to descend.

The flexion of the head may and generally does occur before the os uteri is fully dilated. When this is completely effected the head at once descends with the plane of demi-flexion constantly synclitic with the successive artificial planes of the pelvic cavity. As it descends, it simultaneously rotates upon its axis, the occipital protuberance coming nearer and nearer to the median line in front and the bi-frontal suture similarly approaching the median line behind. The course of the head is at first downwards, backwards, and inwards, following spirally the course of the axis of the right canal. The backward direction is soon changed to a forward one as it descends, but is important while it lasts. Mechanically speaking, the uterine force is reflected from the pelvic walls so as to guide the head and induce this result. All parts of the pelvic wall share in guiding the head, but the right ilio-sciatic line is especially effective. The right parietal protuberance is constantly in advance of this line, which has therefore a similar action to the rifling in a gun-barrel. The left parietal protuberance is remote from the left ilio-sciatic line, and crosses it during the movement of rotation before it is brought into very close relations with it. The rotation of the head ceases when it reaches the inferior strait, with the parietal protuberances in front of the ischial spines and its antero-posterior diameter in the median line, the plane of demi-flexion being completely coincident with the plane of the outlet. The ischial spines, which are the continuation of the ilio-sciatic lines, are usually more projecting than any other part of the latter. The inferior strait is therefore well named, being the narrowest part of the pelvis as well as the end of the double tube.

A slight delay is apt to occur here, during which the movement of flexion is continued, if necessary, until the plane of complete flexion becomes coincident with the plane of the outlet, after which the propulsion of the head is resumed. The subsequent course of the head is through the single tube formed by the soft parts, and might with propriety be set apart as a distinct stage of labor—the perineal stage —since a new force is here called into operation.

Before describing it, I will call attention to a few

points in which the foregoing account differs from
the received teaching upon this subject.   Hodge,
whose exposition of the mechanism of labor is the
most complete extant, states* that the central portion of the child's head describes in its descent the
axis of the general pelvic cavity.   This axis extends
centrally through the pelvis downwards and backwards (afterwards forward), following the curve of
the sacrum.   The axis of the right ·canal, in which
it is here asserted that the centre of the child's head
moves, extends spirally downwards, backwards, and
inwards.   If a piece of card-board be cut out, of elliptical outline, similar to the outline of the occipito-
frontal plane, or the plane of demi-flexion, and applied to the pelvic inlet, so that its long diameter
corresponds to the oblique diameter as already described, the centre of the ellipse will be found to be
at quite an appreciable distance to the right of the
median line.   But if the ellipse is placed in the pelvic outlet in the same manner as the head occupies it
during labor, its long diameter, and therefore its centre, will be exactly in the median line.   Therefore in
moving from the superior to the inferior strait the
centre of the head moves towards the median line, or
inwards, as does the axis of the right canal.

As a necessary concomitant or preliminary to this
inaccuracy of the existing doctrine, some vagueness
of expression concerning the true position of the

* System, p. 30.

head at the inlet will be found, for if the latter had been accurately noted, it would at once have been manifest that the head does not occupy the inlet centrally. The same author states,* "In the first position of the vertex, after flexion has been perfected, it is strictly correct to say that the nape of the neck, or sub-occipital region, is opposite the left acetabulum, and the anterior fontanelle to the right sacro-iliac symphysis; while the right parietal protuberance is to the right acetabulum and the left to the left sacro-iliac symphysis." These four points, the parietal protuberances, occipital protuberance, and anterior fontanelle, are about equidistant. A head which has its occipital protuberance opposite one acetabulum and its right parietal protuberance opposite the other acetabulum, would, if finished upon the same magnificent scale, be difficult to place in the human pelvis. The correct position is stated on pages 46--47.

The importance of accurate discrimination in these points will be more apparent in connection with the application of the forceps. It is sufficient to note here that the head, not being placed centrally, leaves quite a large free space in front of the left sacro-iliac symphysis.

As the head passes through the inferior strait, and even a little before, it begins to encounter the resistance of the pelvic floor, against which it is pro-

* Op. Cit., p. 148.

pelled. This brings to bear upon it the force described at page 37. Assuming that complete flexion has taken place at the outlet, as is customary, the plane of complete flexion is coincident with that of the outlet. As the head is propelled forward in the line of the resultant of the two forces, the plane of complete flexion continues to maintain its coincidence with the successive transverse planes of the parturient passage. The flexion of the head is, however, not kept up, but extension occurs progressively during the remainder of its course.

The movement of extension is readily seen to be somewhat different in its results from the mere reversal of flexion. This is due to the different circumstances under which the movements take place. Flexion at the inlet resulted in bringing new planes of the head in relation with the pelvic planes, and the same is true throughout the pelvis. But the extension which occurs after the passage of the inferior strait has no such displacing effect, the cervico-bregmatic diameter continuing to coincide with the antero-posterior diameter of each successive plane of the passage. Extension occurs because of the great curvature of the canal at this point, which takes a direction almost opposite to that of the bony canal. This necessitates a bending of the projectile upon itself, since the body cannot at once be dragged down with the head. This movement keeps the smallest attainable outline of the head in relation with the vaginal tube. The sub-occipital region remains under the sub-pubic arch, while the forehead and face sweep over the perineum. The perineum becomes greatly distended and changes its shape. It is, as before noted, wedge-shaped or triangular upon section, the apex of the triangle being at the verge of the anus. As the head glides upon and over it the apex of the triangle moves forward and a large portion of the anterior wall of the rectum is added to the perineal surface.

It is very necessary to remember this forward motion of the perineum in any attempts to assist the natural mechanism. As the head escapes from the vulvar orifice the perineal tissues retract to nearly their original condition, chiefly by reason of their inherent elasticity, aided somewhat by the action of the transverse muscles of the perineum. The vulva will then embrace the child's neck, while the head, released from the tube, is again flexed. So far as the forceps are concerned, we might here suspend the account of the mechanism of labor, but for the sake of completeness and for the light which may be thrown on the foregoing stages, we will continue it. At the moment of birth the head was propelled almost vertically upwards (the woman being upon her back). while the body remains behind and in a general way at right angles to the long diameter of the foetal head. Hence the flexion or dropping of the chin when the head is born. A lateral movement is also

described, called restitution, in which the head turns
obliquely after birth, with the occiput in front and
to the left, as when at the inlet.

This is of little importance, nor does it always occur, since it depends upon the manner in which
the body conforms to the mechanism by which the
head was delivered. As the head passes the inferior
strait the shoulders enter the pelvis if the neck is of
its ordinary length. As already noted, their proper
method of entrance is with their long or bis-acromial
diameter coincident with the left oblique diameter of
the inlet, and their elliptical outline in connection
with the beginning of the left canal. This is the
natural provision ; after which they descend in that
canal, rotating in the oppositedirection to that which
the head followed. After the delivery of the head
they arrive at the inferior strait with their long diameter in the median line and the right shoulder in
front. Circumstances cause this mechanism to be
often varied from. The mobility of the neck and its
varying length do not render it absolutely necessary
that the shoulders should follow the rotatory movements of the head or be affected by them. *Per con-
tra*, the shoulders may be prematurely and unduly
influenced by the head rotation. Hence, when the
head has assumed its directly antero-posterior position at the inferior strait the shoulders may have
been compelled to engage in the inlet with their long
diameter directly transverse and thus out of relation

with either canal. Since they have not the solid and comparatively unyielding organization of the head, there is less need for their conforming strictly to the requirements of the passage, and they may, under these circumstances, be dragged or pushed through the pelvis, without any reference to the separate canals, until they reach the inferior strait. Here the bis-acromial diameter will prove too long, under any ordinary compression, to pass through the strait in coincidence with the transverse diameter of the strait, and the shoulders must rotate as they would have if they had started right in the first place. It will be to a great extent a matter of accident whether they rotate so as to bring the right shoulder in advance, as it would have been after descent in the left canal, or the left shoulder, as would occur after the descent in the right canal.

But if the former occurs, the back of the child being directed to the left side, the free head will have its occiput turned towards the left, and in the latter case, the child's back being to the right the occiput will also turn towards the right. It is not, therefore, proper to say that observance of the direction in which the movement of restitution is made will show us what the original position of the head at the inlet must have been. Very generally the shoulders observe the natural mechanism and the bis-acromial diameter becomes coincident with the left oblique diameter of the inlet with the right shoulder in advance. If the outline of the shoulders is not unduly large this relative position of shoulder and pelvic outline is maintained until complete delivery. A plane passing transversely through the shoulders continues to be synclitic with the successive planes of the parturient passage until at the vulvar outlet it is expelled. The right shoulder remains stationary at the sub-pubic arch, while the left shoulder sweeps over the perineum. Where the shoulders are a little larger than common, the plane just mentioned becomes oblique from the moulding of the shoulders, so that the left or posterior shoulder is crowded in advance of the right shoulder and maintains this position throughout, arriving at and passing through the vulvar outlet before the right shoulder instead of simultaneously escaping. Or it may happen that in the moulding process the right or anterior shoulder obtains precedence.

Opinions differ as to which of the two is the natural course, and probably from a want of sufficiently numerous and accurate observations. Where it is desirable to have exact knowledge, as when we attempt to aid the process artificially, there are reasons for preferring the prior delivery of the left or posterior shoulder. Such an occasion often presents itself. The delivery of the head is frequently followed by a more or less temporary cessation of uterine contractions. Under such circumstances the child may be in danger of asphyxia from pressure upon the funis, if the body is large or the funis wrapped around the neck, so that an immediate delivery of the shoulders by the physician is to be recommended. If the posterior shoulder is made to keep in advance, a shorter diameter than the bis-acromial is permitted to coincide with the antero-posterior diameter of the tube, and a smaller outline being presented the perineum is less distended. This is true whichever shoulder is in advance, but the posterior is usually more accessible to the finger and more easily drawn down. Also, if the posterior shoulder is first delivered, the sharp projection of the shoulder is made to pass over the perineum before the full bulk of the body becomes engaged with it, and is therefore less likely to make a rent in that structure, as so often happens. The rest of the body follows the shoulders at once, being too small as a rule to bear any definite relation to the pelvis. Occasionally the breech is large enough to fit quite closely when, being of similar outline to the shoulders, it observes the same mechanism.

To recapitulate. The head in the first position of the vertex enters the pelvis with its occipito-frontal plane coincident with the initial plane of the right canal, and therefore oblique to the plane of the conjugate diameter. Its *first* movement is a compound lateral and forward flexion, which brings the plane of demi-flexion in coincidence, not with the initial plane of the right canal, but with a plane paral-

lel to that of the conjugate diameter, while at the
same time its outline is diminished.   Its *second*
movement is rotation during descent, the former
bringing the occiput gradually in front while the
centre of the head moves spirally in the axis of the
right canal.   At the inferior strait the flexion is, if
necessary, continued until, if not before, the plane
of complete flexion is made to coincide with the
plane of the outlet, the occipital end being directly
in front.   This relative position continues while the
head undergoes a *third* movement, of extension, during the rest of its course, being expelled from the
vulvar outlet in a state of complete extension, but
with the cervico-bregmatic diameter still at right
angles to the axis of the tube.   Next the shoulders,
having engaged in the left canal, rotate as they descend ; arrive at the inferior strait with the right
shoulder in front, which is detained under the pubes
until the posterior shoulder sweeps over the perineum,
and so out, when the rest of the child promptly
emerges.   During the perineal stage the head moves
in a direction almost completely the reverse of its direction at starting.

This mechanism may be clinically verified in many
cases.   At the outset of labor, when the os uteri is
but partially dilated, and the bag of waters uni-
formed, and the head resting loosely at the inlet, a
careful examination will show it to be situated as
follows : The posterior fontanelle will be almost in-

accessible, being at or above the ilio-pectineal line, opposite a point in front of the left acetabulum. The right branch of the lambdoidal suture will also be difficult to reach, extending from the posterior fontanelle in a direction nearly parallel to the top of the os pubis, and ending in the small fontanelle at the postero-inferior angle of the parietal bone. If the head is still oblique this fontanelle can be felt, and if the head is not unduly large even the ear may also be detected in its neighborhood. But if the contractions of the uterus have already forced the head into a parallelism with the plane of the conjugate diameter they will be entirely out of reach of an ordinary examination at this stage of the labor. The sagittal suture will be felt extending first downwards from the posterior fontanelle and then obliquely backwards towards the right sacro-iliac symphysis, thus having the same general trend as the long diameter of the initial plane of the right canal. The right parietal protuberance will be felt at or below the level of the pectineal line opposite a point to the right of the pubic spine and in a line which, vertically drawn, would pass through the obturator foramen near its inner edge. So far as the finger can determine, the central part of the presentation is midway between the parietal protuberance and the sagittal suture or thereabouts. And yet from the description it is evident that the centre of the presenting part must lie in the sagittal suture and not to one

side.   This apparent discrepancy is due to the curvature of the pelvis, so that the horizon of examination, as we may call the limit of the area within
reach of the fingers, differs from the horizon actually
present at the brim.   The arguments as to the position of the head, based upon the location of the caput
succedaneum which forms during the arrest of the
head at the inlet, are of doubtful value.

Carefully observed and recorded instances are
wanting, as is admitted by Matthews Duncan ; and,
until we have more exact facts, reasoning upon theoretical principles is fallacious.   If I might venture a
hypothesis, it would be that the caput succedaneum
forms in front of the centre of the presentation for
reasons similar to those which cause the anterior lip
of the womb to become œdematous in preference to
any other part of the cervical rim.

As soon as synclitism takes place, the right branch
of the lambdoidal suture ascends above the os pubis,
becoming inaccessible until flexion and the descent of
the occiput bring it again within reach.   This may
happen synchronously, in which case it does not
ascend, but in either case its direction will be changed
and it will no longer be parallel with the top of the
os pubis.   The posterior fontanelle becomes more and
more accessible with each degree of flexion.   As rotation and descent proceed it becomes more centrally
situated, being nearer the median line as well as lower
in the pelvis.   The left branch of the lambdoidal suture

becomes apparent as soon as the head begins to rotate, and even before, to some extent, when the head is well flexed. The right parietal protuberance recedes almost directly backwards and to the right side, and when the head has reached the inferior strait each protuberance may be felt with some difficulty exactly opposite to each other, while the posterior fontanelle occupies the median line ; in the centre, if the head is completely flexed, a little above or in front, if flexion is less complete. At this time the sagittal suture extends directly backwards, the two branches of the lambdoidal suture extending from it above like the arms of the letter Y. This adjustment of parts to the pelvic tube is continued throughout the remainder of the labor. The occipital protuberance, being in advance of the fontanelle, appears first at the vulva, and as the latter orifice is enlarged the rest of the presentation is gradually uncovered until the parietal protuberances are exposed, when the head slips out.

The compound flexion, rotation, and extension are easily observed and verified in the succession of events, but the inward motion of the head is difficult if not impossible to appreciate by direct observation. The distance travelled is short, especially in front, where our observation is mainly directed. The occiput also rotates in an opposite direction to the course which the centre of the head travels, which further obscures the problem. But although it cannot be

directly traced with the finger, it is evident enough
from the conformation of the pelvis, and receives further corroborative proof during the use of the forceps.

Variations from this mechanism may and do occasionally occur, and are of some practical importance.
They may be said to consist in either an exaggeration
or deficiency of some of the natural processes. Thus,
a want of sufficient flexion at the inlet may cause a
long delay at the inferior strait, while this defect is
being remedied, or the head may fail to engage at all,
for the same reason. Flexion may be too great, or
rather extension may fail to occur at the proper
point, causing delay in the perineal stage. A misap-
plication of the propulsive force may interfere with
rotation, or the head, being unusually small, may descend obliquely throughout, and even be born in that
manner. Other variations arise from a disproportion
between the head and pelvis, from a want of elastic
force in the perineum, or from other organic causes.
But where the head and pelvis are each normal and
proportionate there is seldom any deviation from
the above-described process.

The time occupied in the movement of the head
through the pelvis varies in the same individual even,
from different circumstances. Normally, in multi-
paræ, ten or fifteen minutes suffice, after full dilatation of the os, to complete the delivery of the child.
In primiparæ, from a half hour to an hour and a

half is usually required, one half of which time is
consumed in the perineal stage.   Where the propulsive force is of ordinary strength these limits are
rarely exceeded, and if they should be in any case,
the cause for the delay should be carefully determined and if possible removed.   The amount of delay which should be regarded as demanding instrumental interference will be discussed in a subsequent
chapter.

2. The Second or Right Occipito-Anterior
Position (R. O. A.) of the vertex is less frequent
than the first, for reasons already assigned, occurring
perhaps in ten per cent. of all positions of the vertex.
It is possible that it is frequently only a stage of the
third position, as will be mentioned under that head,
which was the view taken by Naegele in all cases.   It
theoretically offers more difficulties and is more apt
to need assistance than the first position, from the
comparative smallness of the left canal and the en-
croachment of the rectum.   So far as my own observations extend this is perceptibly true, but the difference is not great.   It follows precisely the same
mechanism as the first, with its direction of motion
reversed, and the description of the former mechanism will answer as well for it, substituting throughout the account " right " for or " left " wherever
needed.   In this position the occipito-frontal plane
coincides with the initial plane of the left canal at the
beginning of labor.   The head then descends in that

canal, its centre following the axis of the left canal
until the point of fusion at the inferior strait is
reached, when it proceeds in the same course and
manner as the first does, during the remainder of its
course.   Where there is any difference in the mechanism, it usually consists in a longer delay at the
inlet at the beginning until flexion is absolutely complete.   The shoulders descend in the right canal as a
rule, and are more apt to observe a uniform mechanism than in the first position because they are naturally placed in the more roomy canal.   These two
occipito-anterior positions are the only ones in which
labor can strictly be called normal.   The pelvis is
evidently constructed with a special design for such
a mechanism, and although other positions and
presentations have often an uncomplicated and easy
termination, they all have some elements which are
apt to give trouble and which show that they are exceptional.

3. In the Third or Right Occipito-Posterior
Position of the Vertex (R. O. P.) the head occupies the right canal as in the first position, but with
its occipital end reversed.   It is more frequent than
the second position, for some of the same reasons
which determine the prevalence of the first, and occurs in about seventeen per cent. of all positions of
the vertex.   At first sight there appears to be no
reason why the same mechanism will not answer for
both anterior and posterior positions.   If the calibre

of the tube is elliptical in outline it might be supposed that the similarly elliptical outline of the head
might descend in whichever way the ends might
point, whether in front or behind.

For several reasons, however, the mechanism is
quite different. The principal cause of this is to be
found in the manner in which the head is joined to
the body, the point of attachment being towards the
occipital end, instead of in the centre of the head.
This causes the propulsive force, which is transmitted through the vertebral column, to act in a line
too far back in the pelvis. The parietal protuberances are also placed on the wrong side of the ilio-
sciatic line. To which we may add that the occipital
end of the head is, if not larger, at least more firm and
resistant than the anterior. The effect of these conditions will be best understood by observing the
course of the head.

Four methods of delivery are possible in this position ; and yet, in spite of this variety, nature is often
incompetent to complete the task. The method
usually regarded as the most common one is as follows :

a, First, Mechanism.—At the beginning of labor
the head is placed with the occipital protuberance
opposite the right sacro-iliac symphysis ; the anterior
fontanelle opposite a point in front of the left acetabulum ; the left parietal protuberance is in front of the
beginning of the right ilio-ischiatic line, and in close

relation with it; the right protuberance is just to the left of the sacral promontory. The occipito-frontal plane is coincident with the initial plane of the right canal, so that the head is obliquely placed as in the first position, but in the contrary direction, the left side of the head being lower than the right. The first effect of the uterine contraction is, as before, to remove this obliquity and bring the occipito-frontal plane into parallelism with the plane of the conjugate diameter. Flexion is also coincidently instituted, but with a modification of its effect. If the occipital extremity of the head impinges closely against the right sacro-*sciatic* arch, which is usually the case, flexion has a tendency to bring the vertebral column of the child still further backwards in the pelvis and to wedge the head in the chord of the arch—*i.e.*, in the right sacro-cotyloid diameter (CD in Fig. 20). The bi-parietal diameter is too large to be so disposed of, and therefore the resistance of the ends of the arch, viz., the promontory and an opposite point in the right ileo-pectineal line, throw the head forward. Flexion has, in itself and apart from the direction of the force, a tendency to throw the bi-parietal diameter forward and nearer the central line, and this operates also to make the head clear the narrow space in which its occiput would otherwise be detained.

If a comparison is made between the outline of the head and pelvis it becomes apparent that without this forward movement of the head there would be a permanent arrest at this point, since the bi-parietal diameter would lie, not in the left oblique diameter as in the first position of the vertex, but in the chord of the sacro-*sciatic* arch, which is always smaller than this diameter. The disadvantage of having the propulsive force transmitted so far back in the pelvis is therefore considerable. Flexion having continued until this difficulty is obviated, the head descends in the right canal with a spiral rotation in the axis of that canal, the occiput becoming more and more posteriorly situated, until it nears the inferior strait. At this level it encounters such an outline as is represented in Fig. 18, in which A and B

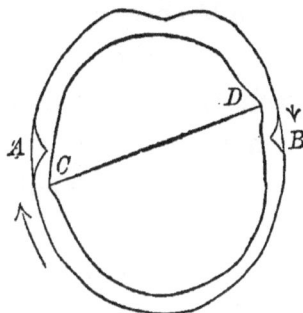

Fig. 18.

mark the position of the ischial spines, and the oblique line CD the bi-parietal diameter of the head. The arrows show the direction in which it is rotating. Now, at either end of this diameter are the parietal protuberances, and to complete posterior rotation and bring the occiput fairly in the sacral concavity, the protuberances must ride over the ischial spines or the ilio-ischial lines just above them. This is not feasible if the proportions between the head and pelvis are at all close. Therefore, the bi-parietal diameter must beat a retreat and occupy the position it takes in the second position at this stage, where, from the fact that the ischial spines are back of the central meridian of the pelvis, only one of the protuberances has to cross the ilio-ischial line, and that not in a close relation. In other words, although the canals are nearly identical here, there must be a transfer of the head from the right to the left canal.

Since an ellipse cannot be turned within its own circumference, flexion must persist until the circular outline of the cervico-bregmatic plane has been reached, and then it is possible for the head to rotate from the right to the left canal. In so doing, the previous motion is simply reversed and rotation continued until the occiput is brought in front and the head placed precisely as in a right-occipito-anterior position of the vertex after it has reached the inferior strait. This is accomplished mainly by the action of the shoulders. The elliptical outline of the shoulders was found to have its long diameter at right angles with that of the head. If in Fig. 19 the long diameter of the shoulders, AB, is placed over the biparietal diameter, CD, where it actually falls, its ends would project decidedly beyond C and D; therefore, in applying such an outline to that of the inlet, Fig.

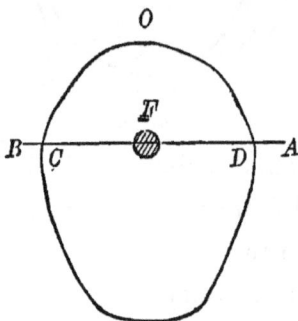

FIG. 19.

20, the shoulders will be evidently seen to extend beyond the limits of the chord of the arch CD. And if the bi-parietal diameter was too small to remain in that relation, much more will the shoulders be deflected elsewhere.

In the great majority of instances the shoulders will be thrown to the right of the vertebral column, since the right shoulder will impinge upon the vertebral column just above the promontory of the sacrum. They are therefore forced to enter the right canal with the back of the child antero-laterally placed instead of entering the left canal, which at first sight appears more natural. This brings the long diameter of the shoulders parallel to the antero-posterior diameter of the head while the latter is rotating posteriorly about half way between the inlet and outlet, and the neck is thereby twisted through an arc of 90°. This involves tension of the neck, and therefore the development of an untwisting force, which becomes constantly greater, for as the head attempts to rotate posteriorly, the shoulders being stationary at the brim will cause it to be resisted, and as soon as the head offers a circular and turnable outline, the untwisting

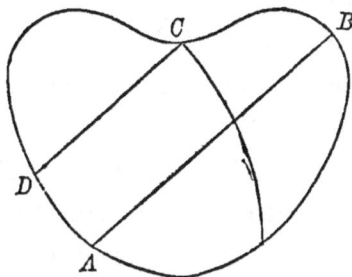

FIG. 20.

A B, line in which the shoulders fall in anterior position.
C D, line in which the shoulders fall in posterior position.

force, added to the uterine efforts, accomplishes anterior rotation, and the head enters the inferior strait
with the occiput in front. It is probable that the
oblique direction in which the uterine force is transmitted tends to promote rotation at all times. This
is at least worthy of investigation. A slight varia-
tion of this mechanism is occasionally observed in
which posterior rotation does not continue until so
low a level as the inferior strait, but flexion is either
completed at the brim or completed synchronously
with descent, anterior rotation also occurring gradually throughout. By this commingling of the steps
the head is already rotated anteriorly or nearly so,
by the time the head arrives at the inferior strait.

b, The second mechanism consists in anterior rotation of the occiput at the inlet and an immediate
conversion into a second position (R. O. A.), at that
point. It is generally believed that the first mechanism is the most common, but, as already stated, Naegele attributed all second positions of the vertex to
this second method. With existing data it is impossible either to prove or disprove the allegation, and
hence we may properly classify the positions as when
we first see them, otherwise this would be the most
frequent method of delivery in this position, the R.
O. A. being entirely discarded. Its occurrence is
favored by the large size of the head and a delay in
flexion. In such case, the disproportion will be too
great to allow the occiput to descend at all while pos-

teriorly placed, and it is therefore forced anteriorly in the only direction in which it can enter. The position of the shoulders has also much influence upon it. If the child, *in utero,* is so placed that its back looks to the right side of the mother, it is obviously a matter of indifference whether the occiput is turned in front or behind ; but if the child's back is turned directly forward the occiput must of necessity come forward also, sooner or later, if there is much resistance at the beginning of its descent. The forward turning of the body, when originally placed nearly in the antero-posterior line, may be due to the uterine contractions, voluntary movements of the fœtus, or a change of position of the woman, which involves pressure of the abdominal muscles upon the child through the uterine walls. It is the most favorable mechanism and the one to be brought about artificially, if possible. After its accomplishment the head proceeds as in the second position.

c, The third mechanism consists in continuous posterior rotation, the occiput remaining posterior throughout the whole delivery. Where there is a great want of correspondence between the head and pelvis, due to the smallness of the former or largeness of the latter, the head may descend with the occiput posteriorly or in any other way, like a shot in a musket-barrel. But in cases where a more exact proportion exists, a definite and distinct mechanism is observed. The head descends in the right canal as in

the first mechanism, until it reaches the level of the
ischial spines when, instead of anterior rotation occurring, the occiput rotates posteriorly. This is effected by great compression and moulding of the head,
so as to diminish the prominence of the parietal protuberances. If the shoulders are placed transversely
at or above the inlet, with the back squarely to the
mother's back, posterior rotation must occur or none
at all. The head is therefore arrested and moulded
by the propulsive force until its bi-parietal diameter
is sufficiently reduced. Usually complete flexion
first occurs with an abortive attempt at effecting the
first mechanism. When the ilio-sciatic lines or the
parietal protuberances are of average prominence
this is a tedious performance, consuming much time,
strength, and patience ; neither are the natural efforts always adequate. When posterior rotation is
complete the head is placed in the outlet with the
cervico-bregmatic plane coincident with the plane of
the latter and the occiput directly posterior. The
disadvantages of the position accumulate as it attempts to proceed. The manner in which the vertebral column is attached to the head causes the propulsive force to be transmitted behind the centre of
the head and pelvis alike. The greater the flexion,
the nearer the foramen magnum is to the occipital
end of the head, and hence the line of force transmitted by the vertebral column to the condyle on
each side of the foramen is thrown backwards by

flexion. The head is therefore forced against the end of the sacrum, or at best against the base of the coccyx, and the secondary force originating in the pelvic floor cannot so well reach the head to impel it forward. The uterine force must then be spent in moulding the head until it is long enough to reach to and be affected by the perineum. The occiput remaining stationary the head is cylindrically moulded so that the cervico-bregmatic plane is thrown in advance of the outlet and a new plane made to take its place, not by an extension of the head, but by its being compressed into a longer shape. After a time, if the head is compressible, and the force holds out, the head becomes long enough to be acted on by the perineal force, and is then conducted to the vulvar outlet and expelled. Where from the small size of the head and body this moulding is unnecessary, the cervico-bregmatic plane continues to occupy the same position as in the case of the L. O. A., but with the occiput behind, and is so expelled, the forehead gliding under the sub-pubic arch. The perineum is in more danger of laceration from this mechanism than from any other; since the propulsive force is directed so far back upon it, that it may be said to attack it in the rear. The occiput is also more pointed than the forehead, and more apt to make a rent during its transit. This is, then, an unnatural mechanism, even when spontaneous, and is to be prevented if possible.

d, A fourth termination exists, rarely witnessed, but which may be taken advantage of in some cases to the great benefit of the perineum.   In this the mechanism is precisely the same as in the third method, until the head is completely beyond the inferior strait, and resting on the perineum with the anterior fontanelle within the lips of the vulva.   At this point anterior rotation may take place, the head rotating around the axis of the cervico-bregmatic diameter, from left to right, until the sub-occipital region is brought under the symphysis pubis.   It is then an occipito-anterior position, and is expelled as such. This was noticed to occur spontaneously by Cazeaux* in one instance, and in another I have brought it about by manipulation.†   It likewise is probably due to the influence of the neck and shoulders.   If the child's back is directed anteriorly, the untwisting force of the neck may be resisted while the head is in the bony pelvis, but whenever it has escaped from it into a tube which is dilatable in more than one direction, this force becomes irresistible, and whirls the head around with the occiput in front.   Even when the shoulders descend with the back posteriorly, the untwisting force may be considerable after they have advanced to any extent in the pelvis, though rarely enough to effect anterior rotation.

---

* "Midwifery."   Edition 1869, p. 367.
† "American Journal of Medical Science," January, 1877.

Clinically, the third position of the vertex may be observed as follows : At the beginning of labor the anterior fontanelle, or its posterior edge, may be felt in front of the left acetabulum, or about in the same position as the posterior fontanelle occupies in the first position. It may usually, but not always, be distinguished from the latter by its large size and quadrilateral shape. The sagittal suture is found extending diagonally to the right in the same fashion as in the first position. There is also a suture extending from the anterior fontanelle corresponding to the right branch of the lambdoidal suture, viz., the coronal, but it is more accessible at the beginning of labor, though also nearly parallel to the top of the os pubis. The left parietal protuberance is to be felt just in front of the right acetabulum, being much further back than the right one is in the first position. The bi-frontal suture is sometimes regarded as a means of diagnosis in this position, since when it is felt we may know that there are four sutures radiating from the fontanelle. It is scarcely ever to be felt, however, and not at all unless the head is abnormally extended, and the ear can, under ordinary circumstances, be felt with less difficulty. The horizon of examination is similarly limited as in the first position, the centre being at a point near the anterior end of the left parietal bone, where the caput succedaneum forms, if at all.

The most important distinction between the first

and third positions as regards diagnosis is in the effects of flexion.   In the first position the posterior fontanelle becomes more and more accessible during its progress and during rotation and descent, and finally occupies a central position.   In the third, the anterior fontanelle, which has the same relative position, is raised by flexion, and while at the inlet recedes in direct proportion to its degree.   If complete flexion occurs, the anterior fontanelle entirely disappears and the posterior fontanelle may be felt behind and to the right of the centre of the horizon of examination.   During descent the anterior fontanelle is never centrally placed, even when in the median line after complete posterior rotation.   If the head is small and flexion incomplete it may be felt in front during the whole of the third mechanism, but otherwise if the plane of complete flexion comes to be at right angles to the axis of the canal, the anterior fontanelle is not felt after the beginning of the labor until birth.

4. The Fourth or Left Occipito-Posterior Position of the Vertex (L. O. P.) bears the same relation to the third that the second does to the first, having the same mechanism in delivery, but with the direction of motion reversed.   It occurs in not more than three per cent. of all positions of the vertex, but the same possibility exists here as in the third, that a few first positions were originally in the fourth, and rotated at an early stage at far above the inlet.

For as the third is converted into the second by anterior rotation, so the fourth is converted into the first by the same movements and under the same circumstances. Anterior rotation at the inlet is more likely to occur in this than in the third position, on account of the presence of the bowel to the left of the sacral promontory. The smallness of the left canal also favors anterior rotation, and therefore it cannot be said to be more difficult than the third.

At the beginning of labor the occipito-frontal plane coincides with the initial plane of the left canal with the occiput behind, and with a similar substitution of "right" for "left," the description of the third position throughout will answer for this one.

I think we may be justified in drawing the following conclusions concerning occipito-posterior positions of the vertex.

First, they are not strictly natural positions. *Secondly,* they have nevertheless definite mechanisms of delivery which under favorable circumstances are alone sufficient to secure their birth. *Thirdly,* if the head and pelvis are of average size, their spontaneous delivery is attended with considerable delay in the labor and may be altogether impracticable. *Fourthly,* in a large proportion of cases the safety both of the mother and child will be promoted by artificial delivery. *Fifthly,* as the pelvic canals are of elliptical outline, the head cannot turn so as to be placed in an occipito-anterior position until it presents a circular plane whose diameter corresponds with the shortest diameter of the pelvic canal. *Flexion* is therefore the first requisite in all methods, whether natural or artificial. *Sixthly*, to make anterior rotation feasible, and with safety to the child, the shoulders must present with the back anteriorly, or be so rotated if they are not so originally. *Seventhly*, failing this, the forceps will greatly assist in the requisite compression, and also enable the physician to control the passage of the head over the perineum more effectually.

II. The Face Presentation.—In the facial end of the cranium there is described a plane called the trachelo-bregmatic, which is named after its long diameter, which passes from the anterior border of the anterior fontanelle to the front of the neck. The transverse diameter of the plane, the bi-malar, measures about three inches, the long, or trachelo-bregmatic, about three and a half inches ; it is therefore somewhat elliptical in outline. It is nearly parallel to the cervico-bregmatic plane, but a little smaller in its circumference. This is the plane which in this presentation corresponds in most particulars to the occipito-frontal in the vertical positions ; entering either canal, and in two ways, with the lower end or chin in front or behind. The chin, then, or *mentum*, takes the place of the occiput in the nomenclature of

these positions, which are as follows : 1, Left mento-
anterior ; 2, right mento-anterior ; 3, right mento-
posterior ; 4, left-mento-posterior. This is also, as
near as may be, the order of their frequency, which
is not great in any position, since the face is said to
present only once in two hundred and fifty or three
hundred labors. They are supposed to occur as the
result of displacement of the vertex, either from
wrongly directed force due to some mechanical difficulty, or as the result of voluntary motion on the
part of the child. I have seen two cases in which a
shock to the mother a day or two before labor was at
least followed by a face presentation. In one, the
house in which the woman was, was struck by light-
ning two days before labor came on, and the sudden
start which one would naturally make under such
circumstances may very well account for the displacement. In the other, a large picture fell from
the wall upon the mother's head, having, no doubt,
a similar effect.

1. LEFT MENTO-ANTERIOR POSITION.—If a head
is placed at the inlet in the third vertical or right
occipito-posterior position, and moved well back
under the right sacro-iliac arch, the parietal protuberances may be made to impinge upon the sacral promontory behind and the ilio-pectineal line in
front, while the occiput also rests upon the brim.
This, as before shown, is usually resisted by flexion.
But should the head nevertheless become impacted,

the propulsive force acting through the vertebral column of the child can move only the long arm of the lever, since the short one is wedged fast. Hence, extension will occur, and as it takes place the head again becomes free. If the extension proceeds further than is required to bring the occipito-frontal plane below the level of the plane of the conjugate diameter, the line of force is thrown in front of the *foramen magnum* with increasing effect, and the extension is accelerated. This continues until the trachelo-bregmatic plane takes the place of the occipito-frontal, with the chin in front of the left acetabulum, and the anterior fontanelle opposite the right sacro-iliac symphysis. This, however brought about, constitutes the first position of the face presentation. The comparatively small size of the trachelo-bregmatic plane makes it unimportant as well as uncertain whether there is any lateral obliquity or not. For the same reason we may say that the face *per se* offers no difficulties in delivery, and if the head was disconnected from the body it would at once descend to the level of the cervico-bregmatic plane ; after which the mechanism of delivery would be the same as in a vertical position. It is the manner in which the body and neck are made to enter the pelvis that constitutes the chief obstacle in its delivery.

If the head and pelvis are of average size, the head descends in the right canal with its trachelo-

bregmatic plane coincident with the successive planes
of the pelvic cavity. But as it descends, and the
head approaches the inferior strait, the body, or
rather neck, is drawn into the pelvis. This brings
the length of the antero-posterior diameter of the
neck to be added to the depth of the cranium, or in
effect to the cervico-bregmatic diameter. This is
too much for the pelvis to accommodate, and the
head must be flexed to remove this difficulty. Flexion would have to occur in any event, if the head is
to advance. But the line of force is through a point
in advance of the centre of the head, and has no tendency to bring about flexion. This can only be accomplished by the action of the secondary or perineal force, and the head is too high up to be reached
by its influence. The further the unaltered head
moves under the propulsion of the uterine force, the
greater is the difficulty, since the occiput and neck
become more firmly impacted in the posterior part of
the pelvis. The only resource of nature is to mould
the head, by which process it becomes long enough
to reach where it can be pushed forward and flexed
by the perineal force. The neck is also liable to be
compressed and moulded, from which great danger to
the child arises, for the neck is ill-adapted for such
pressure, and the circulation in the foetal brain is
much interfered with. When the head is able to be
flexed the difficulty is mainly over, the mere act of
flexion causing the head to sweep over the perineum

and to bring the face to the vulvar outlet, unless the
neck is unusually short.   During descent the head
rotates, so that the chin appears in front and the
anterior fontanelle behind, but this rotation is not
due to any correspondence of the trachelo-bregmatic
plane to the pelvic canal.   This plane is too small to
bring this about, but the cervico-bregmatic plane
which follows it is the one which regulates the mechanism.   For this reason rotation is not as early in
the facial positions, not occurring until the head has
descended well in the pelvis, and the last named
plane become engaged.   It is then regular and complete, the head emerging from the vulva with the
chin under the symphysis pubis.   It escapes power-
fully flexed, and is immediately extended again, after
birth.   Where the child is not large this mechanism
is almost as natural as the corresponding vertical position, the trachelo-bregmatic plane simply preceding instead of following the cervico-bregmatic plane.
In fact, if the head alone were concerned, it would
be a more favorable position than the third vertical
position.   But the implication of the neck and chest
make it a dangerous one for the child, and tedious
for the mother if delay is necessary to mould the
head.

   Its clinical recognition is sufficiently easy at the
beginning of labor, but if it is delayed at any point,
and especially if it is detained at the inferior strait,
this may become a little difficult.   The tissues of the

face allow of a caput succedaneum, or swelling, to take place much more easily than the scalp, and the face may be greatly puffed up and distorted from this cause. Although as a general thing this swelling subsides soon after labor, there is always a risk of irreparable damage to the eyes, and lesser injuries. Delay in delivery is therefore to be deprecated, and should not be permitted to anything like the extent which would be allowed in another presentation.

2. THE SECOND, OR RIGHT MENTO-ANTERIOR, is similarly produced by the extension of a head originally in the fourth or left occipito-posterior position of the vertex. It has precisely the same mechanism as the first facial position, with the direction of motion reversed. It descends in the left canal, having at the beginning of labor the chin in front of the right acetabulum, and the anterior fontanelle opposite the left sacro-iliac symphysis. It does not need to be more particularly described.

3. THE THIRD, OR RIGHT MENTO-POSTERIOR POSITION, is produced by extension from the first vertical position. But it will at once be noticed that it cannot be produced in precisely the same way as a mento-anterior position. There is no chance for the wedging of the bi-parietal diameter in front. The extension of the head must, therefore, be attributed to other causes. Barnes's theory of too great friction anteriorly may be tenable if coupled with an anterior diversion of the uterine force, but after all

there is nothing more probable than the cause assigned by Hodge, viz., the muscular movements of
the child itself.   It is well that it is not easily
brought about, since it is especially difficult and dangerous.   In   this   position   the   trachelo-bregmatic
plane enters the right canal, as in the first facial position, but with its long diameter reversed.   The
chin is found opposite the right sacro-iliac symphysis,
and the anterior fontanelle in front of the left acetabulum.   Difficulties begin early.   In all positions
the anterior part of the presentation moves less rapidly than the posterior, because of the curved con-
struction of the pelvic canals.   Hence in this case
it happens that the forehead remains at the brim
while the chin and base of the cranium essay to advance along the posterior part of the pelvis. This tends
to bring the neck and chest at once into the pelvis,
and the obstruction begins at once.   For this means
that a diameter of seven inches attempts to crowd into
one of five inches in length. The head would naturally
tend to rotate posteriorly with the chin to the rear,
until a new influence is felt in the descent of the
shoulders.   In the mento-anterior positions the
shoulders follow in the opposite canal from that in
which the head descends, but in this, as in the occipito-posterior positions, the shoulders are thrown
back over the right sacro-iliac arch.   The shorter
the neck and the speedier the impact of the shoulders the better, for the left or posterior shoulder is

thrown to the right of the sacral promontory, and the shoulders are thus brought over the entrance to the right canal. This causes the chin to rotate anteriorly, and converts the position into a second or right mento-anterior position, when it is finished, as in that case. This anterior rotation may occur at the inlet, but may also take place between it and the inferior strait. It is closely analogous to the first mechanism of the third vertical position, and is the most favorable one in this position. If anterior rotation does not occur, and the shoulders enter the left canal with the back in front, the chin rotates into the median line posteriorly and the head becomes intensely extended. No relief is afforded even when the head is permitted to reach the inferior strait, since flexion cannot occur, nor could it assist if it did, and extension has already reached its limit. The further the body descends the tighter the wedging ; and anterior rotation, the only resource, becomes more and more difficult. Under any circumstances there must be a great delay until the head is so moulded as to be born in this fashion, and if it is at all large this is impossible. The face will also be fearfully swollen and the head extremely " wire-drawn." The necessity for aid, either manual or instrumental, therefore, to promote rotation at an early stage, is clear.

4. THE FOURTH, OR LEFT MENTO-POSTERIOR POSITION, has the same mechanism as the third, with

the direction of motion reversed, and is therefore
sufficiently described in the above account, with due
substitution of " right " for " left."

The treatment of the facial positions will be considered incidentally in treating of the applicability
of the forceps to such cases.

PART II.

THE FORCEPS.

INTRODUCTION.

"SIR," replied Dr. Slop, "it would astonish you to know what improvements we have made of late years in all branches of obstetrical knowledge, but particularly in that one single point of the safe and expeditious extraction of the fœtus, which has received such lights, that for my part (holding up his hands) I declare I wonder how the world has—"

"I wish," quoth my uncle Toby, "you had seen what prodigious armies we had in Flanders."—STERNE.

"To procure easy travails of women, the intention is to bring down the child, whereunto they say the load-stone helpeth; but the best help is to stay the coming down too fast."—BACON.

WHEN Lord Bacon penned this sage remark, the forceps were unknown, and in the light of other days we are reminded of the fox and the grapes, and similar instances of the depreciation of the unattainable. For almost ever since their rude beginning in the instrument of Chamberlen there have been many who shared in the views of Dr. Slop, as to the blessings of the forceps. When we think of what the instrument can do, and of the numberless lives which it has saved, it is difficult to avoid his enthusiasm, and yet it must be confessed that he was aptly answered. For there is a debit as well as a credit side, and it needs little research to learn that the forceps have also been chargeable with much harm, so that in many minds even now the balance

is doubtful concerning them. That the fault is not
in the forceps, but in the users, it will be my endeavor to show. They are not simply a pair of
tongs, to be applied—somehow—to the child, and
pulled upon—somehow—until it is dragged out, but
a carefully adapted instrument, intended to be applied in a definite way and used in a definite manner, according to the case in which they are used.
And when used with understanding, and under
proper conditions, they fully justify all the eulogy
which has ever been bestowed upon them.

The obstetric forceps are composed of two separate and similar pieces of steel, each of which is fash-
ioned into a blade and handle. The pieces are made
to cross each other near their middle, or at the junction of the blade and handle, at which point a de-
vice known as a lock is contrived so that compression
of the handle will cause an approximation of the

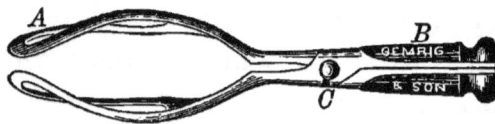

FIG. 21.—DAVIS FORCEPS (UPPER VIEW). A, the blades;
B, the handles; C, the lock.

blades. They are, as has been well said, a pair of
steel hands, to be placed one on each side of the
child's head, to grasp it and draw it from the
mother. Like hands, too, they can grasp lightly or
forcibly. They are intended, primarily, to deliver

a living child from an uninjured mother. But they can also be used to squeeze and drag down a dead child, in the place of craniotomy. Whether this is ever proper is another question.

FIG. 22.—DAVIS FORCEPS (SIDE VIEW).

The first idea, then, of the forceps is of a *tractor*, to be used to supplement or supplant the expulsive forces of the mother. To adapt them for further usefulness in conditions of disproportion between the head and pelvis, they are also made capable of compressing the head so as to diminish its diameters, and thus constitute a *compressor*. They may also be used to further the natural mechanism by flexing, extending, and sometimes by rotating the head, and in this sense may be regarded as a *lever*, but any use of the forceps which implies a leverage upon the sides of the obstetric canal, *i.e.*, upon the mother's tissues, is unscientific, dangerous, and criminal. A properly constructed forceps will embrace these three functions in one, the form of the instrument being determined by these requirements. It would be interesting to trace historically the successive changes which have been made in the forceps during the two hundred years of their employment, but as this would be of little practical value, it will be better to consider only the ideal forceps as at present adapted.

For convenience we will consider first the blade, which is the part in front of the handles, then the handles, and lastly the lock.

1. THE BLADES.—The blades should be large enough to cover a considerable part of the surface of the head, so as to hold it securely, and with as little pressure as possible on any one part. And since they are frequently demanded, because of the tight fit of the head in the pelvis, they must not take up any additional room by adding to the diameter of the presenting plane of the head. For these reasons the blades should be wide, but with a large fenestrum, through which the parietal protuberances of the head project. In this way they will not add a fraction to the size of the head. If the blades are narrow they will not exert so equable a pressure upon the head. Also in this case the fenestrum will be correspondingly small, and the convexity of the head cannot so well protrude between the branches of the blade ; the diameter is therefore liable to be increased by such an instrument. A good, wide blade, with a correspondingly wide fenestrum, is the first requisite in the forceps. It is alleged by some that a wide blade is more difficult to introduce than a narrow one—which is in a measure true, but since the wide blade can always be readily introduced in any case which is suitable for the application of the instrument, it is of

no consequence that another blade can be more readily used. A blade only a finger-breadth wide could
be introduced still more easily, but would be of no
use. A width of two to two and one-eighth inches
will be sufficient, with a fenestrum one and one-half
inches in breadth.

a, Head Curve.—When a pair of scissors, for instance, is opened, the points widely diverge, so that
an instrument made in this way with straight blades
would have a very slight grasping power. In fact,
the only hold which such blades would have upon an
object would be such as powerful lateral pressure
would give. This in the forceps would be a great objection, since the object to be grasped is the more or
less compressible head of a living child, and such pressure is liable to injuriously affect the intra-cranial
circulation, if not the integrity of the brain itself.
Compression of the head is at times desirable and
necessary, but in many, if not most, instances, all
that is required of the instrument is that it shall hold
the head securely with a minimum of compression.
For this reason the blades are bowed outwardly to conform to the curvature of the head. This is known
as the *head-curve* of the forceps. It must not be so
slight that the head will readily slip from between the
blades, nor must it be very great, else there would be
great difficulty in applying them. With a proper
head-curve the tips of the blades will approximate to
such an extent, when the instrument is applied, that

traction upon the blades brings their distal end upon the farther end of the head, so as to not only securely hold it, but also to push it onwards. When forceps are said to slip during their use, one of two things is certain ; either the head-curve of the instrument is insufficient, or the blades have not been properly applied. In the Davis forceps the tips of the blades are one-half inch apart when the instrument is closed, and when open sufficiently to hold a head measuring four and a half inches in the biparietal, the tips are two and three-quarter inches apart. It is obvious that if the head is really in the blades of this instrument, they cannot slip unless the steel blades are so thin as to allow of their being sprung outwardly at the tips. This latter is an accident which I think does occasionally happen in some forceps, but is guarded against in the Davis forceps by a secondary head curve in the blades, namely, a curving from above downwards. This twisting of the blades makes them much stronger, for the outward acting force of the child's head is applied almost edgewise to the arms of the blades, instead of through their thinnest diameter. This secondary curve also adapts the instrument more exactly to the convexity of the head. The forceps are also held upon the head by the pressure upon them of the soft parts and pelvic walls, and in cases where there is not a tight fit and the forceps are applied merely for lack of uterine contractions, an instrument with no head-curve at all may be sufficient to withdraw the head. In difficult cases the head-curve is absolutely necessary, and in any event, the instrument which is useful only in the cases where it is least needed is not a desirable one.

b, Pelvic Curve.—The curvature of the pelvic tube in its whole length is considerable. As before shown, the child in escaping from the vulvar outlet takes a direction almost exactly opposite to that in which it enters the pelvis. Much of this curvature is indeed in the soft parts, and therefore both variable and capable of being overcome by a straightening pressure against the walls. It is true that a pair of forceps which are nearly straight *quoad* their length can be made to seize the head when quite high up in the pelvis and even at the inlet, but it is much more convenient to have the instrument conform to the curvature of the pelvis. This is known as the pelvic curve, and is surprisingly different in different instruments, varying from a barely perceptible curve to one in which the ends of the blades are nearly at right angles to the rest of the instrument. The curvature of the male catheter, for instance, is comparatively uniform, and there is no reason why so great a diversity should exist in the forceps in this respect. The pelvic curve of the Davis forceps, which is greater than that of most instruments, seems to me to be most suitable. It not only enables us to apply the blades to the head at any point with great facility, but it allows them to be adapted to the head in a superior manner. The blades, by reason of this curve, will be more nearly parallel to the axis of the presenting plane of the child's head than if the blades were straighter, and it will therefore be easier to make the traction in the proper direction.

There are, then, two valid reasons for a considerable pelvic curve ; first, that it allows of greater ease in application, and second, that the blades will be applied to the head in a more desirable way. Such an instrument can be used at any point, the straight forceps only when the head is at the inferior strait, without great pressure upon the perineum, and consequent discomfort to the mother. The exact manner of curvature, whether it shall be gradual from the lock to the tip, or whether it shall begin at some distance in front of the lock, is a matter of some moment.

In the Davis forceps (Fig. 21), the shanks of the blades are continued in front of the lock, straight, parallel, and close together, for an inch and a half before either the pelvic or head-curve begins. This insures that the lock shall be outside of the vulva in most cases, even when they are used at the inlet. In many instruments, both curves begin at the lock, which seems to me to be a disadvantage, since the blades are relatively shorter and are unnecessarily wide in the immediate neighborhood of the lock. The pelvic curve of the Davis forceps is peculiar, being increased by widening the fenestrum posteriorly, or rather by having the two bars of the blade nearly parallel, and making the curvature principally in the lower bar. This gives them an exceptionally graceful appearance, which can be appreciated better by direct inspection of the instrument than by any description.

The blades are by these curves fully adapted for seizing and securely holding the head, but they must have handles to facilitate their introduction and to assist in traction, while to admit of compression they are made to cross each other at the lock.

2. THE HANDLES.—The handles are continuous with the blades, and are made of even more diverse patterns than the latter. Some are made of great length, in order to increase the leverage power of the instrument. Some are provided with rings or shoulders, to enable more powerful traction to be made with them. Some are provided with a blunt hook at the extremity, for the same purpose, and for convenience of having a double instrument. Some are made in pieces, so that the handles can be made either long or short. I again select the Davis forceps as possessing the most desirable handles. They are not bulky, are straight and uncomplicated, are long enough to allow of being comfortably grasped by one hand, or even by two, if that were ever necessary. Their length is between four and five inches behind the lock, which is enough. In speaking of traction dur-

ing the use of the forceps, I will explain why I do not regard the handles as the important agent in producing traction, and also show that the length above mentioned is sufficient for the proper use of the leverage power of the instrument. If I am correct in stating that such handles are sufficient for all practical purposes, their advantage over other forms is obvious. They are small and convenient, and there is nothing about them to get out of order or in the way. The rings and shoulders and blunt hooks are in the way during introduction, and have the additional disadvantage of inviting us to make traction in the wrong direction. In some forceps the handles are entirely of steel, and are usually so small that a firm grasp becomes painful to the operator. It is better to affix to them pieces of wood or hard rubber. It is hardly necessary to add that both the blades and handles should have all sharp edges removed, and carefully rounded so as to avoid injury to the tender structures about which they are used.

3. THE LOCK.—The lock is by no means the least important part of the forceps. There are three forms of lock in common use : the English, or mortise-lock, the screw and slot-lock, with which the Hodge forceps is usually provided, and the flat button-lock. The first is the most objectionable. In the first place, the danger of pinching the maternal tissues when the lock is close to the vulva is greater than in any other. Secondly, the forceps may be

locked when the blades are not in exact apposition, but when one blade is introduced a little further than the other. But as soon as traction is made the blades will slide into their proper relation, in which case the blade which has been in advance will usually injuriously scrape the child's head and either bruise or lacerate it. This can be avoided if sufficient care is taken, but it is better to have a lock which utterly prevents it. Thirdly, the blades fitted with this lock cannot at a glance be distinguished one from the other, but must be fitted together before we can tell which is the right and which the left blade. This may appear to be a trivial matter, but any one who has used all kinds will appreciate it. In the second form of the lock one blade is provided with a slot, and the other with a pivot which terminates in a large upright screw-head. When the blades are opposite each other, upon the head, the pivot is just opposite the slot, and may be pushed into it. The screw-head is then rotated between the thumb and finger until the lock is made fast.

There are two objections to this form. First, the looseness of the lock allows of the pivot being inserted into the slot when the blades are introduced to an equal length, but before they are exactly opposite, and when they are somewhat tilted. Then, when the screw is tightened, they may be forced into exactness. Secondly, the projecting screw-head is often in the way, and when the lock is close to the vulva cannot be turned with ease. These objections can be mainly overcome by screwing down the pivot so as to make a close fit before beginning to introduce the blades, but there still remains the fact that the projection is too great for convenience. In the third form one blade is provided with a slot and the other with a closely-fitting pivot which is capped by a flat button-like expansion. When the blades are in exact opposition the slotted blade glides under the button and the instrument is locked, but unless the instrument is exactly adjusted this cannot be done. We have, then, in this lock a safeguard against a faulty application, and when the instrument is locked a guarantee that they are properly applied. The pivot is so low as not to be in the way, and the two blades can at once be distinguished from each other. With the last two locks, the slotted blade is known as the female blade, the one with the pivot, the male blade. Otherwise the blades are distinguished as right and left, according to the side of the pelvis to which they are applied, and sometimes anterior and posterior or upper and lower, which are variable terms, for one blade may be in either position according to the case in which they are used. So far as nomenclature is concerned the slot and pivot lock, then, is much more convenient. I will not undertake the invidious task of pointing out the imperfections of the various forceps now in use; but will simply state my belief, founded on the principles above stated, that the Davis forceps provided with the button-lock, as made by J. H. Gemrig, of Philadelphia, from a reliable model, is the best instrument for general use. It is the instrument used by the accomplished Professor Charles D. Meigs, who declared that it was as near perfection as could be attained, and did not attempt to modify it, and has been used for many years by such veteran obstetricians as Ellwood Wilson and Albert H. Smith, of Philadelphia. Care should be taken in procuring the instrument, for those made by several manufacturers are not correctly made, and leave out some of its most important characteristics.

Such forceps as have special forms or modifications for a particular purpose may be briefly noticed in treating of the use of the instrument. There is no doubt that special skill in the use of any double-curved forceps may enable an operator to use it effectively ; the same amount of skill devoted to the Davis forceps will bring a better return. I say nothing of the straight forceps, because it is nearly obsolete, and every text-book bears witness against it ; nor of forceps for use upon the breech, as this application of the instrument is not yet well established.

THE APPLICATION OF THE FORCEPS.

The forceps, being specially designed and adapted for the head, may be applied to it in any of its presentations and positions, and at any point in its course. The indications for their application will be

discussed in another place, so that we will assume a
suitable and uncomplicated case, in which the os
uteri is fully dilated.   Although the head may be arrested in any part of the pelvis, we are practically
seldom called upon to apply the forceps except in
two situations, viz., when the head is at the inlet or
at the outlet of the pelvis.   It is also necessary at
times when the head is resting upon the perineum
and in great measure through the outlet, but as the
tube is single from the outlet onwards, there is no
difference in the application of the instrument.   At
the inlet the conditions are decidedly different, and
the method of using the forceps is likewise different.

I. We will consider, first, the application when the
head is at the inlet and in the first vertical position
(L. O. A.) in a pelvis which is of normal proportions.   Certain preliminaries are requisite.   First,
the forceps are to be taken from their bag, or case,
and placed in a basin of warm water, so that they
shall be of the proper temperature.   Care should be
taken that the blades are not rattled against each
other while handling them, as the clang of steel is a
peculiarly disagreeable sound to those who are about
to be "operated upon."   Next, the woman should
be placed in a proper position : lying upon her back,
transversely in the bed, with the hips brought to the
edge, so that the vulva overhangs the edge and with
the feet placed upon two chairs.   One, or better two,
sheets may be placed over the limbs, so as to avoid

any exposure, but the vulva should be uncovered so
that the operator shall see exactly what he is about.
Right-minded persons will offer no objections to any
necessary procedure, and it is better to wound the
feelings than the pelvic tissues by uncertain manipu-
lations under the bed-clothes which are certain to get
in the way.   The English prefer to apply the forceps
with the woman upon her side, which is much more
difficult and sometimes well-nigh impossible.   As we
can never be sure beforehand of the amount of difficulty we shall encounter, it is best to secure the most
favorable position at the start.   A third chair should
then be placed between the others, upon which the
operator is to sit, and the forceps are to be placed
within reach.   A supply of lard and several towels
complete the equipment.   If there is any doubt as to
the condition of the bladder, a catheter may be passed, but this is sometimes impracticable.   I assume
that the rectum has been emptied by an enema.

Where are the blades to be applied in the first position at the inlet?   There are several reasons for the
unhesitating answer, on the sides of the child's head.
First, they will fit the head better.   *Secondly*, they
will be less likely to injure the head when compressed than when in any other situation.   *Thirdly*,
they will be applied in a very definite relation to the
head, so that when we move them in any direction we
know exactly in what way the presenting plane of
the head will be disturbed.   *Fourthly*, we can, if

necessary, flex or extend the head, or otherwise control its relations much better when the head is
grasped in this fashion.   In fact, flexion of the head
is next to impossible when the forceps are otherwise
applied.   *Fifthly*, the head can be reduced in size
more certainly than in any other way, since the approximation of the blades compresses and reduces the
bi-parietal diameter, while forced flexion of the head
can be made to reduce the antero-posterior diameter,
by substituting the cervico-bregmatic for the occipito-
frontal.   And lastly, the application is no more difficult than any other in the undeformed pelvis.   The
head, in the position under consideration, is quite
closely applied to the pelvic brim upon the right side
of the pelvis and upon the left side in front.   One
part of the head is at some distance from the brim,
viz., the left parietal region, which is opposite the
left sacro-iliac arch.   One blade of the forceps then
can very easily be placed just where we want it, on
the left side of the head, since there is a roomy passage for the blade.   The blade which is to be opposite this one must be insinuated between the right
side of the head and the pelvic rim to which it is
closely adjusted.   But what is true of this latter
blade is true of both in any other method of application.   It is only when the blades are applied to the
sides of the head that even one of them has a place
provided for it, as it were.

The next consideration of importance is, which

blade should be first applied? In answering this we will notice that one side of the head is posterior and remote, namely, the left side ; the other is anterior and near. The fact that one side is more posteriorly placed than the other will decide the question for us, for if the anterior blade was first passed it would be in the way during the application of the second. These questions being settled, the operator sits in front of the vulva, takes up the male blade of the forceps, and thoroughly anoints the part to be introduced and also his right hand. The latter is to be introduced into the vagina as far as the thumb, or until the finger-tips can be placed between the os uteri and the head. Sometimes the introduction of two fingers will be sufficient for this purpose. If so, all the better, but this precaution should never be omitted, lest the blade should pass to the outside of the cervix, when even a slight amount of force may result in great damage to the maternal tissues. The handle of the forceps should be securely grasped in the left hand ; not held like a pen, which for an object of its weight gives an insecure hold, but held firmly so that it can be introduced with precision. A firm hold does not imply a forcible use, but on the contrary, the ability to grade force with entire delicacy. The tip of the blade is then inserted in the vulva resting against the palm or surface of the fingers, while the tip of the handle is held perpendicularly above the middle of the mother's right groin.

Since this blade is to traverse nearly the whole posterior curvature of the pelvis before coming in contact with the head, the pelvic curve of the instrument is to be first considered, and it is to be passed almost exactly as we would pass a male catheter into the male bladder. As the blade glides upwards, the tip of the handle moves almost directly forward, and with little depression until the blade has reached the lower limit of the side of the head. The head curve of the forceps must now be considered, and the blade made to pass around the convexity of the head. As this movement is executed the handle of the forceps is made to approach and cross the median line, and at the same time is rapidly depressed. As the blade continues to be moved onward in the line of its pelvic curve to a certain extent, the motion is somewhat spiral; but the greater part of the motion in this direction is effected during the first movement; hence, during the second, the line of motion of the tip of the handle is almost straightly diagonal from above downwards. When the introduction is complete the blade is in the free space under the left sacro-iliac arch, and applied to the left side of the child's head. The handle will rest against the perineum and will have its face turned somewhat to the left thigh of the mother, its direction being nearly in the axis of the initial plane of the right canal. As soon as the tip of the blade is felt to be between the cervix uteri and the head, the hand or fingers may be withdrawn from the vagina. The introduction of the male blade is almost without exception very easy. If properly directed to its destination it slips into place almost from its own weight. The second blade does not enter quite so readily, but, under ordinary circumstances, its introduction is not difficult. The right hand is freed from its inunction with a towel and takes up the female blade, which with the left hand is then anointed as in the case of the preceding blade. The right side of the head is much nearer than the left. It will therefore usually suffice to introduce two fingers of the left hand as a guide and safeguard before passing this blade. From the proximity of the right side of the head, the head-curve of the blade has to be considered almost synchronously with the pelvic curve. For almost as soon as the blade begins to be introduced it must be curved around the head. It is therefore held nearly at right angles to the median line, with the handle in the line of the mother's left groin, while the tip of the blade is inserted in the vagina, resting against the palmar surface of the fingers. The part to which we desire to apply the blade is almost directly over the right obturator foramen. The handle is therefore at once moved towards the median line, and depressed as soon as it is clear of the mother's left leg, while it is pushed onward at the same time, so that the line of movement is continuously spiral. I have said that the head was closely applied to the rim of the pelvis on the whole

of the right side of the latter ; but if the head is well
flexed the frontal end will not entirely fill up the right
sacro-iliac arch.   Hence there is a tendency in this
blade to slip posteriorly into this opening.   If it
does the blades will not be opposite, but their con-
cavities will both look forward and they will not grasp
the head and cannot be locked.   To avert this we
keep the blade well forward during its introduction,
and this can be promoted by a simple manœuvre.
One finger of the left hand is retained in the vagina
and placed under the upper bar of the blade.   With
this we can push the blade upwards while the right
hand is urging it onward.   The amount of force requisite for the application of the second blade is usually greater than that demanded by the first.   Where
the forceps are applied only on account of uterine inertia, rather than for any detention from disproportion, or where the head is resting above the inlet
rather than engaged in it, there is not a great difference.   The amount of force which is justifiable cannot, of course, be measured, but when the operator
is thoroughly aware of the true relations of the head
to the pelvis, it is never very great.   When the second blade is thoroughly introduced its handle crosses
that of the first and the slot comes just opposite the
pivot and a slight compression of the handles locks
the instrument.   If the parts of the lock do not entirely and easily coincide, we must withdraw the second blade and apply it with more care until it can

be brought in proper relation with the first. When this is accomplished without difficulty, we may be certain that the head is grasped in its bi-parietal diameter, and may proceed to its extraction in the full confidence that we know exactly what has been done. The position of the handles when the forceps are thus applied is instructive. The head being grasped in its bi-parietal diameter, the face of the handles is directed towards the left side.

But we may notice also that the handles are not in the median line, but point decidedly to the left of it. And the higher the position of the head the greater the divergence of the handles from the median line. If the head was centrally placed in the inlet, as stated by Hodge, this would not be the case. But its centre is decidedly to the right of the median line as we have already stated, and therefore the handles occupy this position, which is a clinical proof of the truth of the views herein entertained of the mechanism of labor. The practical bearing of this will be discussed under the head of traction.

It sometimes happens that when both blades have been apparently correctly introduced, the parts of the lock are still too far apart to be united. This is often due to the fact that they have not been introduced far enough. In this case the handles may be taken one in each hand and pressed well against the perineum, when they will usually lock. When the head is above the brim this is always necessary, and when fairly engaged in the inlet the handles are quite close to the perineum when fully applied. Care should be taken in locking the blades not to pinch the vulvar tissues or allow hairs to be entangled in the lock.

The application of the forceps to the sides of the head when at the superior strait—is taught by Dewees, Meigs, and Hodge, and by a small minority of English and continental authors. Even these admit exceptions, and state that the blades cannot always be thus applied, and Dr. Davis himself was sometimes unable to introduce the second blade of his forceps upon the right side of the head. Nearly all, however, admit the advantages of this method, and merely allege its difficulty. In place of it many recommend that the blades shall be passed with reference to the pelvis, one upon each side, in which case the head will be grasped obliquely. The disadvantages of this procedure are mentioned by implication in the enumeration of the advantages of the method already described.

There is a much greater risk of injuring the head, in addition to the less perfect control which is obtained of its movements. I believe that the objection to the cephalic application of the blades is unwarranted and founded upon several erroneous conclusions.

First, there is not a sufficient discrimination made between the application of the forceps in normal and deformed pelves. It is probable that in some cases of pelvic deformity the blades cannot be applied to the sides of the head. But I utterly deny even the difficulty of application in the normal pelvis, except when from the extreme size of the head no method is adequate, as in hydrocephalus. This is an important point, since the rule should not be conformed to the cases of deformed pelves, which are comparatively rare, but to those in which the pelvis is normal, which are much more frequent. Leishmann says explicitly : " Delivery by the long forceps may practically be considered as an operation in which the head is arrested by reason of contraction of the pelvic brim " (Syst. p. 466). Secondly, there is not enough difference made in the manner of introducing the blades. The English have indeed hardly given in their adhesion to the use of the long, double-curved forceps, having shown a tendency to protract the infancy of the instrument in a characteristic way. Thirdly, there is not enough difference made in the manner of introducing the second blade, and it is improperly introduced. The teaching of Baudelocque, Levret, and Cazeaux is substantially the same as that of Leishmann, which is : " This blade may also be passed in the direction of the hollow of the sacrum." Schroeder, p. 177, is more explicit. " He takes the right blade in the right hand, . . . and proceeds in the same way as just described. Both blades are now situated somewhat behind, and in order to lock the forceps, either both or

at least one of the blades must be brought forward ;
in the first head position the right-hand blade." In
other words, one blade is to be passed under the left,
the other under the right sacro-iliac arch, after
which either the right blade is to be brought forward and opposite the other, or both are to be
brought to the sides of the pelvis until they are opposite. It is no wonder that with such directions
the application is difficult.   Barnes, p. 59, says :
" The instrument held in the right hand lies nearly
parallel with the mother's left thigh, or crossing it
with only a slight angle.   The point of the blade is
slipped along the palmar aspect of the fingers in the
vagina, across the shank of the first blade *in situ*,
inside the perineum toward the hollow of the sacrum.
As the point has to describe a helicine curve to get
round the head-globe, and forward in the direction
of Carus's curve, the handle is now depressed and carried backward until the blade lies in the right ilium."   I do not wonder at his abandoning the attempt to apply the forceps to the sides of the head,
if the second blade is passed in this fashion.   But if
it is held at first, not parallel with the mother's
thigh, but at right angles with it, the blade may begin to curve around the head very soon after it enters the vagina, and can be kept in front with little
difficulty.   And so far from it being proper to pass the
second blade under the right sacro-iliac arch, and
then bring it forward, if we are so unfortunate as to

get it in this position, it should at once be withdrawn and the attempt be renewed to pass it properly.   One reason which is given by Barnes, Fauntleroy, and others, for the pelvic application of the blades is that it dispenses with the need for our knowing the position of the head when using the forceps.   But after they are on, it is of great importance that we should know in which canal the head is situated, and whether our efforts are or are not flexing or extending the head, which cannot be done unless we know the position of the head.   To apply the forceps in a haphazard way to the head is a very unscientific procedure, and is not safe even for experts with the forceps, much less for the unskilled and careless, to whom the doctrine that we need not know the position of the head comes with peculiar comfort.   There are occasionally met with cases in which the determination of the position is extremely difficult, but to make these the basis of a rule is not an indication of progress in our science.

The rules here given for the application of the forceps to the first position of the vertex at the inlet apply equally well to the third vertical and to the first and third facial positions.   In other words, whenever the head is in the right canal, the forceps are to be thus applied.   As this embraces the great majority of cases in which they are used, the doctrine of chances would lead us to apply them in this way whenever we were uncertain as to the position of the head. Any uncertainty, however, can usually be cleared up when the hand is introduced as a preliminary to passing the first blade.

When the head is in the second or fourth positions, or in the left canal, the order of applying the blades is reversed. The female blade is first to be introduced and passed under the right sacro-iliac arch. The male blade is then introduced upon the left side and in front in a similar manner to the introduction of the second blade in the first position. But when this is done the forceps cannot at once be locked, since the blade with the pivot will come on top of the blade with the slot. We therefore take hold of each handle, press them well back towards the perineum, and at the same time slip the handle of the female blade over and across the male blade, when the parts of the lock will be in proper relation to one another. This is a slight inconvenience, but by no means as great as that attending the reverse method of introducing the blade, in which case the anterior blade will be decidedly in the way while introducing the posterior one. The manœuvre should be performed with care and gentleness, remembering that the points of the blades are within the uterus, and are partaking of the motion communicated to the handles. When the forceps are applied upon a head in the left canal, the handles will be observed to extend nearly in the axis of the initial plane of that canal, being to the right of the median line and with their face directed to the right.

II. At the outlet. It is occasionally necessary to apply the forceps while the head is between the inlet and outlet of the pelvis, and therefore imperfectly rotated. The application is made in substantially the same way to the sides of the child's head. Such cases are comparatively infrequent. If the head passes the inlet it is rarely detained until it reaches the inferior strait, and has accomplished its rotation. At this point, or when resting upon the perineum, the forceps are most frequently needed. As the sides of the head correspond to the sides of the pelvis, the long diameter of the head being in the median line, the blades will be applied to the opposite sides of the pelvis, in the following manner : Two fingers of the right hand being introduced as a guide, the male blade is taken in the left hand and held at right angles to the median line, with the tip of the blade in the vulva. As soon as the blade reaches the left side of the head the handle is moved spirally downwards, backward, and onward, while the blade curves around the head and onwards into the pelvis. The same procedure with changed hands is repeated with the female blade, when the handle will be found in the median line, but not pressed against the perineum as when the application is made at the inlet. But no matter where the head is, if it has not completely rotated, the application should be made to the side of the head, which cannot be denied to be perfectly feasible at the outlet, whatever may be thought of the higher operation.

III. The forceps are sometimes applied upon the after-coming head, after the delivery of the body and arms of the child. The method is the same as when the head comes first, the body being held as far as possible out of the way by an assistant during their application and use. The usefulness of the forceps in these cases, is, however, questionable. Only under exceptional circumstances can the child live during the time requisite for their application. If, however, manual extraction should fail, it is commonly advised that they should be used, though Schroeder, for example, does not even mention the possibility of their being required. It is worthy of note that Barnes, p. 75, in speaking of their application to the after-coming head at the brim, says : " The head is engaged with its long axis more or less nearly in the transverse diameter of the brim. The blades should grasp it in an oblique diameter, approaching the antero-posterior." If this is difficult when the head comes first it is much more difficult in head-last labors. Neither is it true in any but deformed pelves that the head enters the brim transversely, for it enters either the right or left canal in the same fashion as when it comes first ; except that it is upsidedown. The head should therefore, if possible, be grasped by the forceps in the same way, but the body and neck of the child are so much in the way, that if manual efforts to deliver the head fail, the forceps will rarely succeed, and craniotomy will be the only resource. Meigs taught that the practitioner should always carry the forceps to every case, lest in a breech case the child should die before we could get them. But, highly as I esteem the instrument, I fear that they have saved few lives under such circumstances.

IV. A few general remarks upon the application of the forceps in any case may here be made. First, they should not be introduced during a " pain " or uterine contraction. The passage of the blade through the cervix will often excite a contraction, which speedily subsides if the manipulation is suspended, after which it may be renewed. Secondly, the use of anæsthetics is neither necessary nor advisable. The introduction of the forceps is not painful, or at least no more painful than an ordinary uterine contraction. The sensations of the woman are also an invaluable guide and safeguard during their introduction. If you are causing pain it is probably because you are not passing the blade properly, and the exclamations of the woman will speedily notify you of the fact. When the blades are locked, you are in no danger of pinching the maternal tissues if the locking is painless. But if the woman is anæsthetized you are left entirely to your own discretion. Although a careful and skilful operator will not do any harm with them under any circumstances, it is much better for the beginner to use them upon a thoroughly conscious individual. After they are once applied there is no reason in the operation itself why an anæsthetic should be withheld, though I would still oppose its use, for reasons foreign to the matter in hand, and therefore inappropriate for discussion in these pages. Thirdly, the forceps should never be taken up with the determination to apply and use them " whether or no." The beginner, and indeed the more experienced practitioner, will occasionally attempt to apply them in an unsuitable case. If when he finds that a blade does not go on readily or that the blades cannot be made to lock, he loses his self-control, and dripping with perspiration attempts to force circumstances and the forceps to obey his will, he will surely do damage. Force is never needed in their application. If they are passed in the right direction they will find their place in every suitable case. Gentleness and skill are the needed elements, and never force. If these fail, let the physician send for some one else, since two heads are better than one. Or, if he is remote from assistance, let him suspend his efforts for a while, meditate upon the cause of failure, and try again. Fourthly, if the blades will not lock readily, it is usually the fault of the second blade, which should be taken out and reapplied instead of attempting to force the blades into locking. If, after due trial, the locking is still impossible, both blades may be taken out and reapplied, while the position of the head should again be carefully made out, since a mistake in diagnosis may have been made, or the position itself may have changed, as occasionally happens.

TRACTION.

The forceps having been applied, the next question is, what are we to do with them? Are we to pull the head out by direct traction, or to pry it out by leverage, and shall it be compressed during either of these movements?

The following propositions may be laid down as a starting-point : *First.* If the Davis forceps (or any other having a sufficient pelvic and head-curve), are applied to the sides of a head at the inlet in the first vertex position, the general line of the blades will be parallel to the axis of the presenting plane of the head. *Secondly.* If traction is made in the line of the blades, the distal ends of the blades will press upon the head, and if the latter is movable will push it onwards in the line of the axis of the presenting plane. *Thirdly.* If during traction the line of the blades is kept parallel with the axis of the canal in which the head is placed, the axis of the presenting plane of the head will be kept in coincidence with the axis of the canal in which it moves. This is what takes place in normal labor, and this is what it should be our aim to imitate with the forceps. It ought not to require a mathematical demonstration to show that when the head is kept in this exact relation with the pelvic canal it will move with the least possible expenditure of force. If instead of this the force be so directed as to push or pull it alternately against the sides of the pelvis, more force will be required, unless the laws of mechanics are altered for the benefit of obstetricians. And yet the great majority of obstetric writers recommend that traction be supplemented by leverage, and that the handles of the forceps should be swayed from side to side that the head may be pried out as well as pulled out of the pelvis. From this it may be inferred, however presumptuous the inference may seem, that they do not make traction in the right direction. Barnes asserts that "pure traction is almost an impossibility," and this is true enough, if the usual directions for the use of the forceps are complied with. A few selections from authoritative works will be sufficient to indicate what these directions are, and in what they result. Cazeaux, p. 970, says : "If the head is at the superior strait, we must first draw downwards and backwards as much as possible." But how? "In performing the tractions the right hand is placed near the clams (at the ends), and above the instrument, the left hand in front of the articulation and beneath." Leishmann, p. 460, says : "The handles should be grasped by both hands. . . The force should be applied as nearly as possible in the direction of the axis of that part of the pelvic canal within which the head lies ; and the operator should act by combining steady traction with a swaying motion of the handles from side to side." Playfair, p. 468, says : "When once the blades are locked we may commence our efforts at traction. To do this we lay hold of the handles with the right hand, using only sufficient compression to give a firm grasp of the head and to keep the blades from slipping. The left hand may be advantageously used in assisting and supporting the right during our efforts at extraction, and at a late stage of the operation may be employed in relaxing the perineum when stretched by the head of the child. Traction must always be made in reference to the pelvic axes, being at first backwards, towards the perineum, etc." And so on through obstetric literature. From these meagre directions we learn that we are to pull upon the handles, and at the same time to see that the pulling is in the axis of the pelvis, " as nearly as possible." That it is not possible seems to be quite generally suspected. Hence the direction of Meadows, p. 224, " When using curved forceps, we should pull less with the handles than with the part of the forceps between the handles and blades." In which case we would insensibly do something else than pull, as will be presently alluded to. Hodge also, p. 253, hints that, " While the practitioner always keeps his right hand on the handles, the left may be variously employed, sometimes in front of the shanks, so as to *depress* the whole head toward the coccyx and perineum, then again the fingers may be applied to the head of the child to watch its progress, and eventually to the perineum, so as to prevent mischief from laceration,

etc., at the time of birth." But in the directions for
traction we have only a reiteration of the advice to
make traction and in the pelvic axis, with the leverage superadded.   I have not the slightest doubt that
the practice of Dr. Hodge was superior to his pre-
cepts, and Lusk says : " Many indeed seek to prevent the anterior pressure of the forceps by placing
the left hand upon the lock and using it as a fulcrum
around which rotation is effected."   Although there
is no written precept for this, I have seen the forceps
used rightly by more than one who would have stated,
if asked, that he was following the ordinary rules. But
can we, by pulling upon the handles, cause the head to
move in the right direction ?   In Fig. 23 we have a
representation of the forceps applied to the head at
the inlet.   The line EF indicates in a general way
the line of the blades and also as nearly as can be
shown in an antero-posterior section the first direction in which the head should move.   If we pull
upon the handles, we will pull in the line of the
handles, and every part of the blades as well as of
the head, which is for the moment immovably con-
nected with the instrument, will move in a line parallel to the line of the handles.   Hence even when the
handles are well against the perineum and traction
is made directly downwards, the head will be pulled
against the symphysis pubis.   Practically, it is difficult to avoid elevating the handles somewhat, especially if the force is great, in which case the head will

be more directly and inevitably pulled against the
symphysis. Between the head and symphysis will be
the bladder and cervix uteri, which will suffer more
or less, according to the amount of force employed,

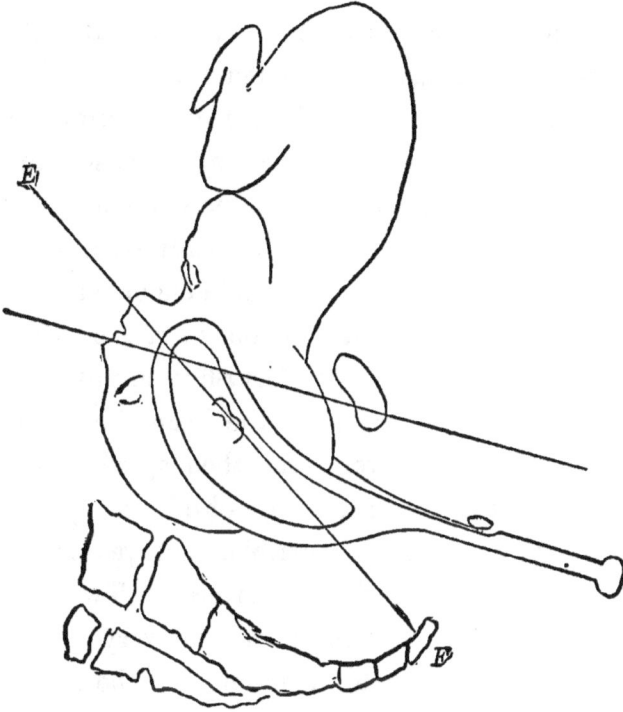

FIG. 23.

while the head will not be advanced. Notwithstand-
ing these facts, there are some teachers who delib-
erately advocate the most powerful traction with both
hands upon the handles of the forceps.

I extract from current medical literature a case
which shows that this teaching is carried into prac-

tice, and that it sometimes accomplishes delivery. The writer has gone to his account. "On the 2d of March, I attended to a Mrs. M., a multipara, third child. The two first were delivered by craniotomy. The vertex presenting R. O. A., and impacted between sacrum and pubes; the conjugate diameter of superior strait greatly contracted. I applied forceps and had considerable difficulty in locking them. Dreading the laceration which might ensue, in this case, from side-to-side lever action, I concluded to rely entirely upon direct and steady traction. My strength giving way, her husband held me around the waist, whilst the patient was held *in situ* on the dorsum, by four women. In forty-five minutes I had the satisfaction of bringing the head down on the perineum. The delivery was then speedily accomplished. Both mother and child, a girl, did well." This is simply horrible, and yet the child was born and the mother recovered. Two circumstances probably determine the delivery when the forceps are used in this manner. In the first place the head finally slides off from the pubes as from an inclined plane. But the amount of force requisite for this is very many times greater than that which would be required if the traction were made in the right direction. In the second place, the head being pressed against the cervix, irritates the uterus into making powerful contractions, which both impel the child in the proper direction and to some extent de-

flect the tractile force of the forceps. The power
which the forceps have in determining uterine contractions by the mere fact of their presence is an important fact, and in many cases greatly diminishes
the amount of force required from the forceps themselves.

Among the first to have a practicable doubt as
to the possibility of making traction upon the handles in the proper direction was Tarnier, who accordingly invented a pair of forceps with a considerable
pelvic curve, which was fitted with steel rods affixed
to the lower edge of the blades,
so that we could pull in the line
of the blades and not in that of
the handles. This is an unnecessarily ingenious contrivance,
since we possess in the ordinary
forceps all that is necessary if we
will use them correctly.

The method which seems to
me to be the correct one, I will
now attempt to describe. When
the forceps are applied at the inlet the handles are seized by the
right hand from above and held
firmly, compressing the head as
little as possible at first. The

Fig. 24.

left hand is placed so that the ball of the thumb
comes over the lock (see Fig. 24), while the index-

finger rests upon the upper arm of one blade, and the middle finger upon the other. Now, while the right hand holds the handles almost at rest, the fingers of the left *push* upon the blades so as to move them and the contained head downwards, backwards and a little to the left of the median line. Secondly, while the fingers are pushing downwards in this way, we may also make use of them as a fulcrum, and by elevating the handles cause the blades to move in an opposite manner, but care must be taken that the force thus applied by the right hand is not enough to overbalance the downward pressure of the left, else we will merely extend the head without propelling it. It is sometimes convenient to vary the position of the left hand and fingers, but the principle is the same, that pushing and not pulling is the first step in traction. When the head begins to descend we may place three fingers between the blades, the thumb and little finger being upon the outside, and combine a pulling with a pushing motion upon the blades. But throughout the handles are simply elevated and not pulled upon, or but slightly, having due regard to the proper direction, and bringing them into the median line only when the head has reached the inferior strait. When the head is delivered the handles will lie upon the abdomen of the mother. This, in brief, is the method which I employ and advise. When we consider the comparatively small amount of force which the fingers can exert, it is in marked contrast to the method of employing the united efforts of two men in pulling upon the handles, and will scarcely be credited with sufficient power by those who are accustomed to use much force. But when we reflect upon the statements of Poppel and Kristeller, that a force of from four to eight pounds is often enough to expel a head that had lain immovable for hours, it is evident that traction in the right direction need not be very forcible. For the forceps are used perhaps oftener for simple uterine inertia than for any other reason, and it is especially in these cases that I recommend this method. And it is also evident that traction which impels the head against the pubes instead of in the proper pelvic axis must always be unnecessarily powerful in every instance. When this method is carefully and patiently carried out, it will rarely fail to deliver if the case is a suitable one for the employment of the forceps. But there are occasionally met with cases in which more force is demanded, in which the method must be modified. In such cases we may pull upon the handles with the right hand and with such force as may or can be exerted, while at the same time we endeavor to deflect the force in the proper direction by pushing upon the blades in front of the lock with the left hand, at the same time making use of the leverage above described. But under no circumstances will it be necessary to pull upon the handles with both hands, or put the

foot against the bed, or secure additional help in
traction. If the force which can be exerted in the
right direction in this way is incompetent to deliver
the child, no amount of force wrongly applied will
succeed without injuring the maternal tissues to an
utterly unjustifiable extent.

Having defined what I mean by traction, the details
of the operation may be enumerated. The tractile
efforts should be made during the continuance of
the labor pains, if the latter are frequent and regular,
and suspended in the interval between them. But as
the pains will rarely be of this character, it is usually
allowable to pay little attention to them. They should,
however, be imitated, with some exaggeration. Traction may be made during one or two minutes, and then
suspended during two or three minutes. There are
several reasons for the intermission. In the first place,
continuous pressure will be undesirable for the mother,
and will weary her. To give rest between the efforts
is therefore necessary. Secondly, it either dilates the
vaginal tissues too rapidly, if we succeed in continuously advancing the head, or it interferes too much
with the circulation in the parts in advance of the
head if the latter does not advance. If the pressure
is intermittent this is avoided. Thirdly, there will
be more or less compression of the head in every case.
If the traction is continuous the compression, whatever its degree, will be continuous and the circulation in the child's brain will be dangerously inter-

fered with as well as that of the parts to which the blades are directly applied. For this reason it is advisable, whenever we have ceased traction temporarily, to partially or wholly unlock the forceps in order to take off all compression exerted by them This is done by sliding the female branch partly or altogether from under the button of the male branch. When we resume traction, the simple grasping the handles relocks the instrument, and allows us to proceed as before.

The whole time occupied in traction varies greatly in different circumstances. In a simple case of uterine inertia without disproportion, the only consideration in the way of immediate delivery is the due preparation of the soft parts. Ten or fifteen minutes is all that is usually required by the multiparous woman for the accomplishment of this part of labor naturally, and we may conform to this in using the forceps. Where there is much disproportion, we may have to wait much longer before we can deliver, during which time the head is moulded, as in protracted labor without the forceps. I do not think that any rule can be laid down as to the longest limit of traction. Ellwood Wilson has kept them on for eight hours (Am. J. Ob., 1876), using them only during the pains and merely to assist the latter. This is an exercise of patience which would overtask most of us, and would not be safe as a rule for general application. The duration of the second stage of labor

for eight hours, with the head well in the pelvis, is
not entirely devoid of danger, under any circumstances, though when we reflect that the compression of the head by the forceps really relieves the
maternal tissues to that extent, it is probable that
labor might be allowed to continue much longer
while the forceps were applied than without them.
When we find that the head does not advance under
our efforts, made in the proper direction and with full
compression, we may decide when to abandon the
forceps for the perforator by the condition of the
mother.   So long as that continues good and the pelvic tissues show no signs of injurious pressure, we
may continue our efforts until thoroughly satisfied
that the head cannot be delivered by the forceps. But
in the vast majority of cases, if the forceps can be applied and locked, they will be competent to deliver,
under an hour.   And it cannot be too often repeated
that there is nothing to be gained by becoming im-
patient and hanging with our whole weight upon the
handles of the instrument.   So long as force is applied in the right direction, any amount which can
be exerted may be employed.   The safeguard is that
a great deal of force cannot be applied in the right
direction, and if it is used in any other direction it
becomes at once unjustifiable, whatever its amount.

When the head is upon the perineum, it is sometimes well to make tractions between the pains, and
not during their continuance.   This applies mainly

when the expulsive efforts are violent, for in that case the added force from the forceps will favor perineal laceration. This plan was first suggested by the late Dr. S. D. Turney, and will sometimes be found useful. There are some who recommend that the forceps should be removed when the perineum has become greatly distended, for fear of laceration.

The forceps give us such a thorough control over the head and its movements that I believe they are a great help to prevent rather than to cause this accident. We can hold back or advance, flex or extend the head with entire ease, as may be needed. But to do this successfully requires coolness, judgment, and quickness, and a wrong turn of the forceps at the critical moment will certainly cause a laceration if this is at all imminent. If a person is not quite sure of himself, he had therefore better take them off rather than wield a power potent for evil as well as good. When they are removed the head may be extracted by the form of rectal manipulation known as the Ritgen-Goodell method, although Smellie (Coll. 19, cases 1-2) described and used it, and gave the credit to Ould. Two fingers are introduced into the rectum and placed upon the forehead of the child, while the thumb of the same hand, or fingers of the other hand, are placed upon the occiput through the vulva. The head is then manœuvred out in a manner easier to perform than to describe. When this is done during the absence of a pain we certainly escape from rupturing the perineum, so far as the head is concerned.

In taking off the forceps when the head is on the perineum we consider mainly the head-curve of the instrument. Having separated the lock, one of the handles is moved across the median line so as to lie in the groin of the opposite side, which will cause the blade to glide out of the vagina without disturbing the head at all. The same is done with the opposite blade in the contrary direction. When for any reason it becomes necessary to remove the forceps at a higher level, the pelvic curve may have to be considered, or in other words the blades are withdrawn in the same manner in which they are applied, with a reversal of direction.

COMPRESSION.

The utility of the forceps as a compressor is beyond question, since the bi-parietal diameter is capable of being diminished by their use from a half inch to an inch. As we can reduce the antero-posterior diameter in another way, we can by compression greatly facilitate delivery. But when the forceps are not applied to the sides of the head we must be very careful how we use compression on the living child. The question in such a case is not so much whether we can diminish the size of the head by compression, but whether we will not cut and injure the head by it. When applied to the sides of the head, and this should include the great majority of cases, compression carefully performed is entirely innocuous and of great benefit. It should in every instance be performed slowly, evenly, and gently, and should be maintained only for a minute or two at most, with an interval of relaxation following. If the head is suddenly squeezed in the forceps, or if the handles are tied together, as the manner of some is, harm will be done as a matter of course. But if effected as stated above, the full compressing power of the instrument may be exerted without injury to the child. I have been surprised to find after the fullest compression, but intermittently applied, that not even a temporary imprint of the blades could be discovered upon the child's head within ten minutes after its delivery. I do not think that anything is ever gained by continuous compression. The head can be moulded to much better advantage, even in the most difficult cases, by systematically intermitting both traction and compression, even when the question of the child's life is not under consideration.

LEVERAGE.

The action of the forceps as a lever may be invoked in some cases, for the purpose of flexing or extending the head, but I hasten to add that it is not to be used to pry out the head by "to and fro" leverage, as is so generally taught. Denman, p. 376, recommended to use the forceps almost

exclusively as a lever. " The first action with them
should therefore be made by bringing the handles,
grasped firmly in one or both hands, to prevent the
instrument from playing upon the head of the child,
slowly toward the pubes, until they come to a full
rest. Having waited a short interval with them in
this situation, the handles must be carried back in
the same slow and steady manner to the perineum,
exerting as they are carried in the different directions, a certain degree of extracting force ; and after
waiting another interval, they are again to be raised
toward the pubes according to the situation of the
handles." As this would only alternately flex and
extend the head, as well as interfere with any right
direction of traction, it is no wonder that Denman
preferred the vectis, about the use of which he appears to have had a more intelligent understanding.
The more modern method is known as the " pendulum leverage" or " lateral oscillations," and consists
in swaying the blades from side to side while making
traction. This is supposed to act on the principle of
the ratchet. One side of the head is brought down
and is expected to stay down while by a reversal of
the instrument the other side is brought to the same
or a lower level, and so on until it is extracted.
Barnes claims to be able to deliver in this way almost without any traction. Even if it were true
that this method of leverage was preferable to traction in the pelvic axis, and advanced the head, it is

pertinent to inquire how this is effected. If it is done at all, it must be by making each side of the pelvis alternately a fulcrum, against which the forceps are pried. As J. Matthews Duncan says, there is no toothed rack in the pelvis. Therefore, when we bring down one side of the head we must press it with great firmness against the pelvic walls if we expect it to retain its position while the other side of the head is being brought down. In other words, the steel blades of the forceps or the parietal protuberances are alternately jammed against the maternal tissues intervening between them and the pelvic walls upon each side whenever this delectable form of leverage is resorted to. And this happens whether the head really is advanced by it or not. That it does not advance the head seems to me to have been so clearly shown by Dr. A. H. Smith that I take the liberty of quoting largely from his paper (Fig. 25). " Let PW and P'W' be the pelvic walls in section made in the plane of the maximum diameter, and of that transverse diameter, the ends of which are grasped by the blades. Let MM' be the

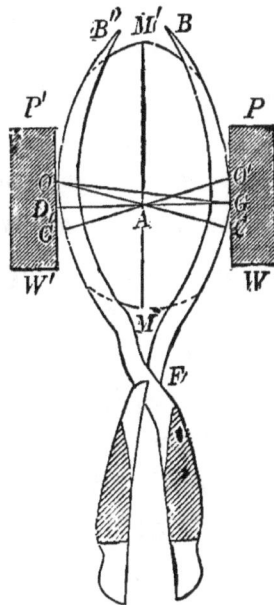

FIG. 25.—AFTER A. H. SMITH.

maximum diameter, corresponding with the axis

of the canal, GD the transverse (whether bi-parietal or other) these two crossing each other at A, which will then be the centre of oscillation of the head in any pendulum movement of the handles and the centre of motion in a direct traction. Let us draw through this centre two oblique diameters, OC, O'C', and also from the extremity of the line GD an oblique line to a point O, on the periphery of the head nearer to M'. FBB' will represent the blades of the forceps through the fenestra of which the tissues of the scalp should protrude sufficiently to rest firmly against the pelvic walls, unless the blades be narrow, when the scalp tissue will come in contact with the pelvis at the sides of the blades. . . . What will be the effect of pure oscillation, or leverage as it is called, with compression, but without traction, the method recommended by Dr. Barnes? The first movement, say, will carry the handles toward W ; the head, then, being ' immovably united to the forceps,' must rotate upon an axis passing through A, perpendicular to the transverse GD, which transverse also rotates, the extremity G moving upward toward P, and the extremity D correspondingly descending toward W'. But as the diameter GD moves, so does the oblique diameter OC, passing through A, move also proportionally ; C following G upward, as O follows D downward, and the extremities of this oblique diameter come to assume the position, in relation to the plane of the pelvis occupied before this lateral move-

ment, by the extremities of the transverse diameter.
But we know that every oblique diameter of an ovoid
passing through the centre of the greatest transverse
diameter is greater than the transverse, and that the
increased length is proportionate to the distance of its
extremities on the periphery from those of the transverse. The more considerably, then, we move the
handles towards W, the more we place the longer diameter of the head in the position originally occupied by the transverse diameter. As the handles
swing back, approaching the median line, the diameter in relation with the plane diminishes until the
handles pass the median, and are made to approach
W' ; when the same change takes place in the bearing
of the extremities of the oblique diameter O'C', and
this diameter takes the place of the transverse against
the pelvic walls. Here, then, we have a demonstration sufficiently clear . . . that oscillation without traction simply brings to press upon the pelvic
walls, with a sort of slow vibratory impact, portions
of the head farther separated from each other than
the points which rested in contact with those walls
before the swaying motion was started ; that while
the pelvic wall is subjected to alternations of excessive pressure and partial relief, there is nothing in
the movement itself to advance the head an iota, the
side which descends with the swaying of the handles
in one direction, ascending equally (unless driven
down by the *vis à tergo*, which acts altogether inde-

pendently of it) when the handles are swayed in the opposite direction.''

In the same manner he demonstrates that "leverage with traction is simply traction plus an aggravation of pressure upon surfaces already so tightly compressed by the circumference of the child's head as to obstruct its advance toward the outlet of the pelvis.'' As I have already shown, the proper direction of traction at first is not in the median line, but somewhat to one side, in the axis of the canal in which the head is placed. With every other lateral oscillation therefore, the head is so far forth impelled by the coincident traction in that axis, which may account for the success in delivery which is claimed for this method. But it is hardly necessary to add that it is not expedient to subject the mother's tissues to pressure for the sake of occasionally making traction in the right direction, when it is equally easy to make it directly and uniformly in a proper manner.

The forceps have a proper use as a lever ; first in flexing the head.

a, Flexion.—A delay in the flexion of the head may be and not infrequently is the sole cause of delay in the advance of the head. We may, in such a case, apply the forceps and by simple traction deliver, but as the occipito-frontal diameter is thus kept coincident with the successive pelvic planes, unless the head is spontaneously flexed in transit by the influence of the pelvic walls, a greater amount of force is required to deliver than if the cervico-bregmatic diameter had been substituted. Preliminary flexion of the head is therefore very desirable. If the head is not flexed the blades of the forceps are not parallel to the occipito-mental diameter of the head when applied, which should be the case when the head is thoroughly flexed. After applying them in such a case, then, before thoroughly locking the instrument, we may elevate the handles. This will allow the blades to glide over the head and become parallel to the occipito-mental diameter. We then slowly and firmly compress the head with the handles in this position, and when the head is thoroughly grasped we return the handles to their original position, pressing against the perineum, at the same time pushing them gently farther into the pelvis in order to slightly lift the head from the brim while the movement is being made. This will flex the head, so that in some cases the amount of force required for extraction will be very slight, if indeed, the uterine efforts are not entirely sufficient. This manœuvre in competent hands is devoid of danger, but the blades must be upon the sides of the head, and we must, of course, accurately know the position of the head before attempting to change it.

In occipito-posterior positions the same principle may be brought to bear with great advantage. The rotation of the occiput forward is promoted by extréme flexion. At the very beginning, we then, may secure this in the following manner : The handles are pressed back firmly against the perineum, as each blade is introduced. The head is then carefully grasped and the handles elevated. Traction is then made with the handles in an elevated position in order to keep the head flexed as much as possible. A similar elevation of the handles is sometimes useful during the perineal stage of an occipito-anterior position in order to extend an unduly flexed head. The application of the same principle in facial positions is sufficiently obvious, as well as in a condition of too great lateral obliquity in the vertex positions. As compression reduces the bi-parietal diameter, and flexion shortens the antero-posterior diameter, the combination of the two procedures decreases the entire circumference of the head.

b, Rotation.—It is also possible to use the forceps to rotate the head, but this application of the instrument is rarely proper or useful. In occipito-anterior positions the pelvic walls will effect rotation much better than we can, if we are careful to make the traction in the right direction. All we need do is to see that we do not hold the handles in such a manner as to interfere with rotation. But in occipito-posterior and mento-posterior positions the desirability of early anterior rotation is so apparent that there is a strong temptation to bring it about with the forceps at all hazards. In these positions, when the head is at the inlet, it is highly improper to attempt anterior rotation with a pair of forceps having a decided pelvic curve. The form of the instrument distinctly prohibits this.

The voice of experience is equally clear against making the attempt with the straight forceps. If then we cannot secure anterior rotation by manipulation, either internal or external or both combined, we may apply the forceps to the sides of the head as it lies and make traction in the axis of the canal in which it is placed without any present reference to its rotation. We should exert as little compressing force upon the head as possible, for this reason. When the head nears the inferior strait it tends to undergo anterior rotation according to a mechanism described in a preceding section. As the parietal protuberances project through the fenestra of the blades the mere presence of the forceps does not interfere with this, and anterior rotation may take place by the head turning inside the blades of the forceps. This has not unfrequently been noticed. Decided compression tends to allow the head to come down without sufficient contact with the pelvic walls to compel rotation, especially if the tractile force is considerable at the same time. The fact of its occurrence will be generally indicated by a tendency of the blades to slide together posteriorly, when the forceps are unlocked in the intervals of traction. For as a uterine contraction comes on, before the forceps

are locked the head, in attempting to rotate, carries
one of the blades with it, leaving the other stationary.   This at least is the explanation I have framed
from observing the phenomena, though it is not entirely adequate.   For under these circumstances it
may happen that as soon as the forceps are unlocked, and when there is no uterine contraction, the posterior edges of the blades at once approximate, which
perhaps shows that during the last traction the head
was prevented from rotating anteriorly by the manner in which the forceps were held ; but as soon as
it is released from their influence it rotates, carrying
one of the blades with it.   If the position of the
blades is not much altered they may be carefully
made to come opposite to each other without with-
drawing them, but if their relative position is much
disturbed, it is an evidence that anterior rotation is
nearly or quite complete, and they may be withdrawn
and re-applied as in an anterior position.   I have
witnessed these changes taking place during the descent of a mento-posterior position, and have re-applied the forceps accordingly.   The innocuousness
of the proceeding was shown by the fact that not
even a temporary imprint of the blades was discover
able upon the head immediately after birth.   If the
head is large it will not rotate within the forceps,
but may rotate with them.   It is in just such cases
that it seems most plausible that we should force anterior rotation with the instrument.   For if it fails

to occur we will have to drag the occiput over the
perineum, or in the case of the mento-posterior position be utterly foiled in the delivery. Nevertheless,
forced rotation will almost invariably prove to be a
meddlesome interference. And although the situation seems to call for the limit of tractile force, we
should also be very chary of this as well.

If traction is very powerful at this juncture, compression will almost certainly be also carried to its
extreme limit, and we may pull the head through the
inferior strait posteriorly and destroy all hope of anterior rotation. What is needed is moderate and
patient traction, and a slight motion of rotation ; so
slight as to be of little service for effecting a change
of position in itself and only to test the inclination
of the head. If the head is manifestly inclined to
rotate it may be gently assisted, but force will do no
good and may do harm. The head and the pelvic
walls between them will determine the exact level
· and time at which rotation can be effected much better than we can, and we should therefore only assist
it when actually being performed and not prematurely urge it. When the head is at the inferior
strait so much of the blades are exterior, that the
intra-vaginal portion of the instrument is sufficiently straight to make it entirely proper to allow the rotation to take place with the forceps applied. But
when it has occurred they should be withdrawn,
when they may be re-applied, or the case left to the

uterine contractions. There are two principal reasons for not attempting to force rotation. In the first place rotation is normally accompanied by descent. The head begins to rotate at the level of the ischial spines, but at the end of the movement may have reached the perineum.

The exact proportion of descent and rotation in a given case is determined by circumstances which we know nothing about in a given case, and not even a skilful operator can cause the head to rotate anteriorly as well as the natural conditions spontaneously bring about. On the contrary he may impede the process by his efforts. In the second place, a certain proportion of cases cannot be rotated anteriorly without twisting the neck of the child to a fatal extent, and doubtfully even then. When the back of the child's body is posteriorly situated in the womb this is true, and this cannot always be known beforehand or remedied by manipulation. To patiently make moderate tractions in such a way as not to interfere with rotation, and to keep the head well flexed, should be our aim in occipito-posterior positions.

WHEN TO USE THE FORCEPS.

The forceps may be used under the following circumstances.

I. For *delay* in the second stage of labor, arising from : *a*, uterine inertia ; *b*, small size of vagina ; *c*, rigidity of maternal tissues ; *d*, obstructions from bands, etc. ; *e*, large size of head ; *f*, want of flexion ; *g*, pelvic deformity.

II. For delay in the first stage occasionally, as in : *a*, placenta previa ; *b*, rigidity of the os uteri ; *c*, absence of a natural dilating agent.

III. For certain accidents of labor, in any stage, and when rapid delivery is indicated, as : *a*, convulsions ; *b*, prolapse of the funis ; *c*, excessive uterine action menacing rupture.

IV. For certain secondary purposes as for : *a*, extraction of the child after rupture of the uterus ; *b*, after gastro-hysterotomy or elytrotomy ; *c*, for removing tumors and foreign bodies from the maternal passages.

The forceps have been and may be used for any of these conditions, though the advisability of their use in a given case must depend upon the individual circumstances then present, and not entirely upon a general rule.

I. It is first in order to define what is meant by delay in the second stage, or what measure of delay calls for the use of the forceps.

When the os uteri has become fully dilated and the liquor amnii has escaped, the great majority of multiparous women are delivered within a few minutes. A second stage of five or ten minutes' duration is very frequently observed and fifteen minutes is probably above the average in normal labor. In

primiparæ, the dilatation of the vagina and perineum usually takes up more time, so that from a half-
hour to an hour is not far from the average in this
class of cases. The length of the first stage has little
to do with that of the second. A first stage of
twenty-four hours may be followed by delivery in ten
minutes, when once the os uteri is dilated, and a first
stage of two hours may be followed by a second stage
of many hours. The second stage may be protracted
from any of the causes mentioned under this head at
the beginning of the section, the most common of
which is uterine inertia, or a want of *sufficient* propulsive power, for the term is a rather relative one.
If the case is protracted beyond the average limit
we may ask ourselves three questions : First, What
harm does the delay do ? Second, Can we safely interfere ? Third, Of what advantage to either the mother
or child will the leaving the case to " Nature " be ?

First. Delay in labor, especially in the second
stage, injures the mother and child in direct proportion to the length of its continuance and the depth
to which the head has descended in the pelvis. Each
expulsive effort is attended with an expenditure of
vital force, while at the same time the functions of
digestion and assimilation are so interfered with that
the drain cannot be kept up indefinitely. The
woman is weaker with each pain. This is provided for in normal labor. The ideal woman during
an ideal pregnancy becomes more robust and vigor-

ous during the whole gestation. She enters upon labor with a reserve of physical force entirely adequate for its performance, so that when delivery is accomplished she may arise, cleanse herself and the baby, and resume the ordinary functions of life with unimpaired vigor. But the ordinary civilized woman with whom we have most to do, finds even an ordinary labor a rather exhausting piece of work, and if it is at all long she requires a proportionately longer time in which to recuperate. Also, the average woman does not only approach labor with a very slight, if any, reserve of physical force, but is too often even below par at this time. Her urine is apt to be albuminous, her blood hydræmic, her digestion impaired, and if she is, under such circumstances, subjected to a long and tedious labor, she is in a ripe condition for all the diseases incident to the puerperal state. So far, then, as the expenditure of vital force is concerned, the sooner the woman is through with her labor the better. She is not only using up her strength by muscular contractions, but she is kept in mental suspense, and is not usually able to repair her energies by the taking of food. The continuance of the second stage also involves the pressure of the head of the child against the soft tissues of the mother, with an increase of the pressure during each pain. This pressure is least when the head is movable at the inlet, but increases in its capacity for evil at least, with each fraction of descent.

Its continuance may result in destroying the vitality of the tissues pressed upon. It is the most common cause of vesico-vaginal and other fistulas and predisposes to the occurrence of pelvic inflammation after labor. When the head is long detained at the inlet the anterior lip of the cervix is apt to become œdematous, which may occur to such an extent as to make it a further impediment to delivery. When the head is long detained at the inferior strait, or on the perineum, the latter structure often becomes boggy and inelastic, and is very apt to become lacerated subsequently. The irritation caused by the pressure of the head upon these structures, which are delicate and amply supplied with nerves, is apt to give rise to convulsions. The child's life is also endangered, especially when the detention is at a low point, for not only is the direct compression harmful but the uterus may grind off the placenta and thus destroy the child. That all these evils follow in the train of delay is universally conceded ; but to come to an agreement upon the time when the danger is imminent rather than prospective is more difficult. Before attempting to fix the danger line we may pass to the second question, '' Can we safely interfere ?'' This depends upon the questioner. If he is ignorant of the anatomy and physiology of the structures involved, of the mechanism of labor, and of the nature of the forceps, he ought not to interfere even by his presence. But any one who is really qualified to attend upon the parturient woman can interfere with perfect safety to mother and child.

The mere application of the forceps contains not a single element which is detrimental, and is not even painful. After they are applied they can hardly be said to touch the mother during traction, since the opposed surfaces of the head project through the fenestra. It is not therefore the forceps, but artificial traction, which is to be found fault with if the operation is objected to. The woman is unable to expel the child, for a want of expulsive power. We supply this power and the woman is delivered speedily instead of waiting indefinitely at great expense of vital force. Statistics are not always reliable, and I refer those who put their trust in them to papers by Ed. S. Dunster[*] and A. M. Fauntleroy,[†] merely citing one specimen. In the Rotunda Hospital, Collins used the forceps only once in 694 cases of labor, with a fœtal mortality of 1 in 26 and a maternal mortality of 1 in 329. Harper used them once in 26 cases, with a fœtal mortality of 1 in 47 and a maternal mortality of 1 in 1490. The average duration of labor was, in the first case, 38 hours, in the second, 16 hours. Barnes (Obst. Oper., p. 280) says : " Properly speaking, the mortality from the forceps is *nil.* Women die because the instrument is used too late."

[*] Proceedings of Michigan State Medical Society, 1878.
[†] " American Journal of Obstetrics," January, 1879.

We gain further light from the answer to our third question, " Of what advantage will it be to leave the case to Nature ?" The usual answer is, that we avoid the dangers of rapid delivery, allow the maternal tissues to be properly " prepared," and lastly, we leave the case in the hands of " Nature," who or which is all-competent and of benign tendencies. It is difficult to deliver with the forceps in less than ten minutes in any case, and thousands of women are naturally delivered in less time. The operation usually excites uterine contractions, and as a matter of fact, post-partum hemorrhage is rare after a forceps delivery, even when they have been applied on account of uterine inertia. And in the matter of preparation, when the labor has continued for an hour or so during the second stage, the tissues will be progressively unprepared and unfitted for delivery the longer it continues. A head stationary in the pelvis, at any point, is progressively congesting and infiltrating the tissues below it, and not preparing them. If it is not stationary, but is advancing with each pain, the pelvic canal is so short that there will be no delay. The truth is, that refuge is taken in a vague appeal to the powers of Nature by those who are too indolent to learn how to render assistance to the mother. In the words of the late Dr. Turney, " It sounds well to talk of trusting to Nature. It is sweetly suggestive of green fields, of flowery meads, of singing birds, of the gentle lullaby

of breeze and falling waters, and brings to mind all
the pleasant sights and sounds which amuse us in a
summer's ramble." But what are the facts? Nature
has ordained that woman shall be safely delivered in a
few hours. The defiance of the laws of Nature for
generations has brought it about that the woman is
unable to deliver herself without undergoing great
danger. And if we were to leave all cases to Nature
a great many women would die undelivered under
this benign *régime.* It is not to Nature that we
leave the woman, it is to the consequences of physical deterioration incurred in defiance of her laws. I
cannot see what advantage there is in this, when we
possess safe and efficient means for rescuing her and
the child from these consequences.

The logical deduction to be drawn from these
premises is, that when the os uteri is fully dilated,
the child should be expelled promptly ; and in the
time observed to be usually consumed in normal labors. If it is not, the longer the labor continues the
more danger the woman and child incur, and con-
versely, the sooner she is delivered by the forceps,
the less risk will they run. While these deductions
are fully warranted by the physical facts involved,
they are subject to modification from certain considerations of a practical character. Many women have
a horror of "instruments" and "operations," and
will be unfavorably agitated by the early suggestion
of their employment. Also, in the existing state of

lay intelligence, if anything whatever should go
wrong with the woman after their employment, the
physician and his forceps will have to shoulder the
blame.   On the other hand, Cazeaux mentions that
the pains of women are sometimes greatly increased
by the statement that the forceps must otherwise be
used.   Having due regard to these considerations,
the following rule seems to me to be proper.

Whenever the second stage of labor has lasted
two hours and the head is still stationary or advancing with great slowness, we should inform the patient
that we are about to apply the forceps.   If we explain the necessity and propriety of the operation we
will rarely find any objections, especially if the
woman is already tired of her fruitless sufferings.
This rule may be deviated from according to the circumstances of each case, but it will more often be
proper to shorten it than to protract the time of
giving relief.   There is no need of keeping the
woman in suffering for hours solely that she may deliver herself, and still less for keeping her under the
noxious influence of an anæsthetic for hours, when
we can safely extract the child at will.

These remarks apply to all cases of delay in the
second stage of labor, but it is necessary to qualify
them in some particulars.   Thus, in obstruction
from cicatrical bands, persistent hymen, and the like,
it may be necessary to incise the obstructing mem-
brane before applying the forceps.   More often, we

may wait until the band is made tense by the pressure of the head within the forceps, against it, before dividing it. A head which is enlarged from hydrocephalus can rarely be delivered by the forceps as well as by a preliminary evacuation of the fluid. But the forceps are useful as an aid to diagnosis in hydrocephalus, since the large size of the head is very clearly demonstrated by the wide divergence of the handles when the blades are applied. And if the head is very large the forcep scannot be applied at all.

In deformities of the pelvis the propriety of applying the forceps has been brought in question, and a few words of justification are in order. The pelvis is rarely deformed throughout its whole extent, the deformity being usually limited to either the outlet or inlet. When the outlet is deformed either by the approximation of the ischia or bending forward of the coccyx, the propriety of using the forceps is unquestioned. But when the deformity is at the inlet and is at all considerable, many prefer version to the forceps. Barnes says (op. cit. p. 244) that the proper range of the operation of turning is from 3.25 " to 3.75 " of the conjugate diameter, at the latter point coming into competition with the forceps. Goodell substitutes version for the forceps when the conjugate diameter is between 2.75 and 3.25 inches. The limit is variously stated by different authors, but is recognized by the great majority as at least equal to the forceps in marked deformities and often succeeding when the latter have failed.

The principles upon which this practice rests were first stated by Simpson. They are, in brief, as follows : First, the transverse diameter of the head can be lessened to a greater degree by the influence of the pelvic walls when the base of the skull is in advance than by the forceps when the head comes first. Second, a greater amount of force can be employed by pulling upon the body and neck of the child, combined with supra-pubic pressure, than by the forceps.

Traction upon the body of the child is capable of greatly compressing the head. Of this there is no doubt. It can be exerted to the extent of producing deep indentations in the parietal bone by pressure against the promontory. But it has its limits. Duncan has shown that on an average the child's neck breaks with a force of 100 pounds and decapitation ensues when the force reaches 120 pounds. We have then a distinct limit to the amount of force which can be exerted by traction after version. The same experimental data are wanting for the forceps, but all the force which they can exert will not affect the integrity of the fœtal structures, and there is every reason to suppose that a force of over 120 pounds can, if necessary, be brought into requisition. The main question is whether it is true that the bi-parietal diameter can be diminished to a greater extent when the base of the skull is in advance.

It is alleged that the base is much narrower than the upper part of the skull, the bi-mastoid diameter being from four to nine lines less than the bi-parietal. Hence, when the vertex comes first, the head tends to flatten out, while when the base comes first, the diameters are progressively diminished during its progress. This is true enough, but we should

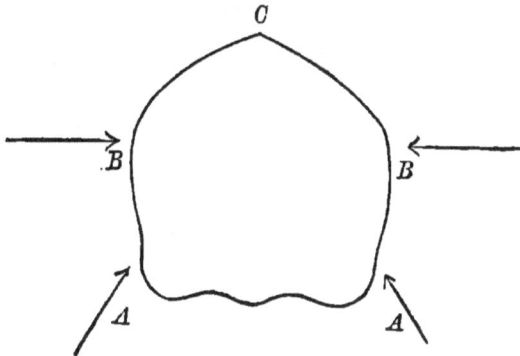

FIG. 26.

contrast the state of affairs in version, not with those obtaining in unassisted vertex labors, but when the forceps are used. Which has not been fairly done. Fig. 26 represents a transverse section of the fœtal cranium. When the base is in advance the compressing force of the pelvic walls will act in the lines indicated by the arrows AA. But when the forceps are applied to the sides of the head they exert their compressive force in the lines BB, or directly. To say that the parietal bones may be made to overlap

at C by forces acting in the lines AA, better than
when acting in the lines BB, is absurd.

Thus we are driven to the conclusion that version
cannot be superior to the forceps, or an elective substitute for it, when it is possible to apply the forceps
to the sides of the child's head. The difficulty of so
applying them has been, I think, greatly exagger-
ated. The deformity usually occurs upon one side
only of the pelvis, *i.e.*, one sacro-iliac symphysis
only has been affected by disease. As the result,
one of the pelvic canals is impaired or destroyed, but

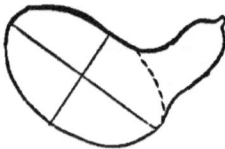

FIG. 27.
AFTER SCHROEDER.

the other is not necessarily interfered with. Such a state of
things is shown in Fig. 27, from
Schroeder. But when the conjugate diameter is reduced to 3
inches or less, both of the canals
are impaired and the normal mechanism is entirely
altered. I admit that the difficulty of grasping the
head in its bi-parietal diameter increases with each
degree of contraction below 3 inches, but we can at
least place them obliquely upon the head in every
instance. When this is done, we can bring to bear
upon the head the compressing force of the pelvic
walls nearly as well as when it is dragged down with
the base in advance, and without the risk of breaking
the child's neck, or any of the unavoidable dangers
attendant upon delivery by the breech. And in addition we will have such compressing and moulding

power as is afforded by the forceps. Nevertheless, if in any case we find it impracticable to apply the forceps to the sides of the head, we would be justified in resorting to version, if the latter were a generally safe procedure. Since head-last labors have a mortality to the child of at least fifty per cent, and since version is attended with decided danger to the mother, especially when performed through a contracted inlet, and also since when it fails we have to resort to craniotomy at a great disadvantage, this cannot be claimed. I cannot then conceive of a case in which version is justifiable as an elective procedure. If it fails, nothing remains but craniotomy. If after applying the forceps we have not enough skill to deliver, then perhaps version may be tried before the last resort.

The manner of using the forceps in a deformed pelvis differs but little from that which is appropriate in the normal pelvis, and that little will be different in each case because scarcely two deformities are exactly alike. One general feature has been pointed out by Barnes, viz., that the promontory of the sacrum usually projects and the head has to make a curved progress around the promontory before it can enter the axis of the pelvis, which he calls the "false curve of the promontory." The effect of this forward jutting of the promontory I conceive to be simply to equally push forward the head and greatly exaggerate the backward direction of the

pelvic axis.   Hence it is often useful to begin our
efforts at extraction in these cases by pushing the
whole instrument downwards and backwards in the
direction of the sacro-coccygeal junction, without
any traction in the ordinary sense of the term.   The
exact nature of the deformity cannot always be made
out at the time of labor, but we can always form a
correct idea as to the direction in which the head
ought to move in order to pass the narrowed inlet.
When this is carefully ascertained, we will find that
a comparatively slight amount of force is often
enough to bring the head past the obstruction, after
which it usually progresses without further hin-
drance.   It is worth while spending any amount of
time in being certain as to the axis in which the head
is to move, for traction in the wrong direction will
be tenfold more useless in a deformed than in a normal pelvis.

II. It is sometimes proper to apply the forceps
during the first stage of labor, or before the os uteri is
fully dilated.   But although we have advanced in
the obstetric art far beyond the point when a delay
of at least six hours upon the perineum was regarded
as an essential prerequisite to their application, a degree of conservatism is necessary upon this point.
For there are some unavoidable dangers attendant
upon their use during the first stage, and the necessity for their employment should evidently counter-
balance these before we resort to them.   These dan-

gers are, first, the possibility of bruising the cervix
during the introduction, which in skilful hands may
be reduced to a minimum ; and, second, the proba-
bility of lacerating the cervix when we come to making traction and cause the head to be pressed against
it. For this there is no avoidance except in imitating
the natural course of labor in making the traction
moderately, intermittently, and patiently, so that the
head may evenly and with as little haste as possible
dilate the cervix before passing through the os. And
yet in the very cases in which the procedure is most
likely to be demanded the cervix is most indisposed
to dilate without laceration.

The fact that a laceration once begun may extend
indefinitely and end in a veritable rupture of the
womb, makes this consideration too important to be
lightly passed over in deciding upon the use of the
forceps in the first stage. The indications which
suggest their employment are as follows : First, long
delay due to the existence of organic rigidity of the
cervix. The most notable case illustrating this use
of the instrument is one reported by Roper, 1874, in
which the cervix was four inches long and as thick
as a man's wrist. After labor had lasted forty hours,
seven incisions were made, and gradual dilatation allowed to proceed for sixteen hours, after which the
forceps were applied and a living child extracted.
This is an extreme case, but the principles of
treatment are the same in lesser degrees of organic

rigidity. First, incision, which should not be deferred so long as in this case ; second, a brief period for further dilatation by the natural powers, and then, or indeed as soon as the forceps can be applied, they may be used to further the dilatation by increasing the force with which the head is pressed against the cervical rim. If incisions are unnecessary, so much the better, but in true organic rigidity they are usually demanded. The greatest care and gentleness is called for during traction, which if at all sudden or violent, would be sure to do harm. In this way we bring a more efficient dilating force to bear against the cervix than in any other possible way. The fact that traction must be made with moderation, and will probably last for some time, is a valid reason for resorting to it early in the labor. We must not wait until the woman is exhausted by her efforts before we begin, or the desperate nature of the circumstances will impel us to work faster than we know to be judicious.

In functional or spasmodic rigidity of the cervix, which has resisted other methods of treatment, it is also allowable to apply the forceps as soon as they can be introduced without violence. There are also certain cases in which the liquor amnii is early evacuated and the head of the child fails to take the place of the bag of waters as the natural dilating agent. In cases of unusual pelvic or uterine obliquity, or when from any cause the head is not forced against the cervix after the evacuation of the liquor amnii, the os fails to dilate. In these cases we will usually find that the cervix is early dilatable although undilated, and if by external or other manipulation the head cannot be made to engage, it is proper to apply the forceps, since otherwise the second stage is not likely to begin.

The duration of labor in the first stage which calls for the application of the forceps varies to a much greater extent than in the second stage. The first stage is extremely variable in length even in the same individual in different labors, and its pains can almost always be endured for a much longer time than those of the second stage. Hence, a duration demanding assistance must be determined in each instance by the condition of the mother. All other approved means are to be tried before resorting to the forceps, but if her condition is at all unfavorable, we should have no concern as to the mere number of hours which have elapsed, but proceed at once to render assistance. Another application of the forceps during the first stage is for the complication of placenta previa. It is sometimes recommended to introduce the forceps after a sufficient amount of dilatation has been secured, merely to cause the head to press against the cervix and so arrest the hemorrhage. This, to be entirely successful, would require the head to be constantly pressed against the cervix. It is much better to first detach the placenta from the cervical zone, after the manner of Barnes, after which the hemorrhage usually ceases. If previous loss of blood and other conditions make it necessary to deliver forthwith, the forceps may then be used, and this application of the instrument is one of the most useful of the modern purposes to which it has been devoted. For without it we must resort to the more formidable operation of version, or await the slow, often fatally slow, spontaneous dilatation of the cervix.

III. Certain accidents of labor require a more or less prompt termination of the labor. In prolapse of the funis, when it cannot be permanently replaced, the forceps may be used in the interests of the child. The forceps blade may be of great utility in itself, in pushing the funis up and out of the way, after which we may make as much or as little traction as is called for, and either promote the delivery with them or allow it to go on without further interference. This will not interfere with the trial of the genu-pectoral posture in replacement. This position has been found useful as a preliminary measure in the application of the forceps in these cases, and is also recommended by Mossmann (Am. Journ. Obst., Jan. 1879), in certain cases of spinal and pelvic deformity. There is such an entire reversal of direction in this position that the operator must know well what he is undertaking ; otherwise it is calculated to be of service.

Convulsions.—The typical puerperal convulsion comes on usually when the head has reached the inferior strait and the bearing down efforts of the mother give rise to cerebral congestion. The indication is then plain to apply the forceps at once and deliver as speedily as possible, administering ether in the meantime, if it is at hand. We thus eliminate one of the causative factors of the eclampsia and generally put an end to the seizures. In the cases which occur during the first stage, rapid delivery is not so necessary. We have ample time to obtain the influence of chloral by the mouth or rectum, bleed, or otherwise control the convulsions according to our lights. Dilatation is usually rapid, and when complete we can apply the forceps with less risk. In cases of hemorrhage before delivery the forceps also afford us the means of promptly terminating the labor.

IV. When rupture of the uterus has taken place, the prevailing practice is to deliver *per vias naturales*, either by the forceps or version if possible. The propriety of this begins to be questioned. First, we will probably enlarge the rent already made. Second, we leave the rent to close spontaneously, which seldom happens. Thirdly, we do not take away what is quite as important should be removed as the child, the blood and fluids which escape at the time of rupture. The elaborate statistics of Dr. Trask show a better percentage for gastrotomy than for ordinary delivery and with the improved

methods of operating now in vogue there is no reason why a much larger percentage should not recover
if we should at once proceed to open the abdomen
after the accident.   The child can be removed, the
rent united by suture, the abdominal cavity thoroughly cleansed from extraneous fluids, and the
woman will be no worse off than after the Cæsarean
section instead of almost uniformly perishing, as
when the abdomen is left unopened.   It is true that
gastrotomy may be performed after delivery *per vias
naturales,* but the latter is an unnecessary step, and
the former would be frequently refused by the
woman or her family if she had been already delivered.   Prevention is better than cure, and the forceps will be found much more useful as a preventive
of rupture.   When the uterine contractions are very
forcible without having any appreciable effect upon
the head, we may justly fear the occurrence of rupture of the uterus.   The exact amount of contraction which justifies interference may be left to the
judgment of the practitioner at the time.   The forceps may also be used for purposes foreign to their
original design.   They may be inserted into the incision made in the Cæsarean section or gastro-ely-
trotomy, in order to grasp the head.   They may be
used to deliver detached fibroid tumors from the vagina, or to extract foreign bodies, such as globe pes-
saries.   But for such purposes the mechanical tact of
the operator in each case is a sufficient guide.

A few words may be added as to the possibilities for harm possessed by the forceps. So far as the mother is concerned we may reiterate the statement that there is nothing in the right use of the instrument which can by any possibility injure her. The animadversions of the earlier writers were due, in part, to their wrongly attributing to the instrument what is the result of delay in labor, and in part to the unavoidable injuries caused by an instrument without a pelvic curve, to say nothing of the heavy, thick, leather-covered blades which formerly belonged to the forceps. But with the modern instrument we can do harm only by violence in introduction, a wrong direction in traction, or by too great haste in completing the delivery. The anterior lip of the cervix has been ground off, the pubic bones have been fractured, the vagina lacerated, by such improper uses ; but none of these things will happen when the forceps are used as herein directed and as common-sense would dictate. And in the normal or but slightly deformed pelvis, it is equally true that the forceps need do no harm to the child, if applied to the sides of the head and used intermittently and judiciously. It must be admitted that even those who are skilled in their use are occasionally mistaken in the diagnosis of the position of the head, and hence apply them over the brow and occiput, but this should not be laid to the charge of the forceps. So, also, when the head is at the inferior strait, but has not completed rotation, the exact state of affairs may be overlooked and the forceps applied obliquely upon the head. In occipito-posterior positions, especially when flexion has not taken place, the ends of the forceps may unavoidably compress important structures, and in deformed pelves indentations of the cranium may be caused by the jutting promontory.

We may have, then, as the result of the forceps, bruising or laceration of the child's scalp, facial paralysis, asphyxia from compression of the medulla, indentations of the cranium. All of these are avoidable by applying the forceps to the sides of the head, except those due to compression of nerve trunks in occipito-posterior positions. Against these we have no safeguard, except the early securing of flexion so as to bring the line of the blades parallel to the occipito-mental diameter, and, failing this, the utmost care in compression and traction, which as already pointed out, is proper for other reasons as well. Direct indentations of the cranium are rarely if ever caused by the direct pressure of the forceps, but the amount of traction necessary to bring a head past a jutting promontory may cause the latter to indent the head, as happens also in head-last labors. The etiology of indentations in general is well worked up by Dr. J. Trush, to whose paper* the reader is referred for a more extensive discussion.

* "American Journal of Obstetrics," July, 1879.

THE END.

A

PRACTICAL TREATISE

ON

SEA-SICKNESS:

ITS

SYMPTOMS, NATURE AND TREATMENT.

BY

GEORGE M. BEARD, A.M., M.D.,

FELLOW OF THE NEW YORK ACADEMY OF MEDICINE; OF THE NEW YORK
ACADEMY OF SCIENCES; VICE-PRESIDENT OF THE AMERICAN ACADEMY
OF MEDICINE; MEMBER OF THE AMERICAN NEUROLOGICAL ASSOCI
ATION; OF THE AMERICAN MEDICAL ASSOCIATION; OF THE
NEW YORK NEUROLOGICAL SOCIETY; AUTHOR OF "NERV-
OUS EXHAUSTION," (NEURASTHENIA); "OUR
HOME PHYSICIAN," ETC

THIS Treatise is not a theory, or dream, but represents extensive experiments of the author, and much
experience at sea, on long and short voyages, and in
different climates.

The philosophy advocated in this work is that Sea-
Sickness is a *functional disease of the central nervous system.*
The treatment proposed is in harmony with the philosophy, and has already been tested, not only by myself,
but by a number of other medical observers, with most
satisfactory results.

The position taken is that sea-sickness, like any other
form of sickness, is an evil to be avoided, and that by
the plan of treatment here proposed it can, in the
majority of cases, be prevented or greatly relieved.

It is designed to make the work clear and practical,
and to adapt it to meet the wants of both practitioners
of medicine and travelers by the sea.

FOURTH THOUSAND—Enlarged by results of experiments.
One 12mo Volume, Extra Cloth, Price $1.

THE NATIONAL HAND BOOK OF
AMERICAN PROGRESS.

A Reference Manual—1492 to the present time.

Edited by Bishop E. O. HAVEN, D.D., LL.D.

Late Chancellor of Syracuse University, N. Y., formerly President of the
North Western University, Illinois, and President of the
Michigan (Ann Arbor) State University.

J. Q. POWER. N. Y.

SIX VOLUMES IN ONE.

HISTORICAL,	FINANCIAL,
BIOGRAPHICAL,	STATISTICAL,
DOCUMENTARY,	POLITICAL, Etc.

Including the **1880 Census,** Population of States and Cities.

The Most Complete and Reliable *non-partisan* record of our Nation's Progress.

CONDITIONS. { Over 500 12mo pages. 60 Engravings. Elegantly Bound. } **$2.00**